SpringerBriefs in Applied Sciences and Technology

SpringerBriefs present concise summaries of cutting-edge research and practical applications across a wide spectrum of fields. Featuring compact volumes of 50 to 125 pages, the series covers a range of content from professional to academic.

Typical publications can be:

- A timely report of state-of-the art methods
- An introduction to or a manual for the application of mathematical or computer techniques
- A bridge between new research results, as published in journal articles
- A snapshot of a hot or emerging topic
- An in-depth case study
- A presentation of core concepts that students must understand in order to make independent contributions

SpringerBriefs are characterized by fast, global electronic dissemination, standard publishing contracts, standardized manuscript preparation and formatting guidelines, and expedited production schedules.

On the one hand, **SpringerBriefs in Applied Sciences and Technology** are devoted to the publication of fundamentals and applications within the different classical engineering disciplines as well as in interdisciplinary fields that recently emerged between these areas. On the other hand, as the boundary separating fundamental research and applied technology is more and more dissolving, this series is particularly open to trans-disciplinary topics between fundamental science and engineering.

Indexed by EI-Compendex, SCOPUS and Springerlink.

Azman Ismail · Fatin Nur Zulkipli ·
Bakhtiar Ariff Baharudin · Andreas Öchsner
Editors

Technological Frontiers and Sustainable Innovations

 Springer

Editors
Azman Ismail
Maritime Engineering Technology
and Centre for Women Advancement
and Leadership
Malaysian Institute of Marine Engineering
Technology
Universiti Kuala Lumpur
Lumut, Perak, Malaysia

Bakhtiar Ariff Baharudin
Maritime Engineering Technology,
Malaysian Institute of Marine Engineering
Technology
Universiti Kuala Lumpur
Lumut, Perak, Malaysia

Fatin Nur Zulkipli
School of Information Science, College
of Computing, Informatics
and Mathematics
Universiti Teknologi MARA
Machang, Kelantan, Malaysia

Andreas Öchsner
Faculty of Mechanical and Systems
Engineering
Esslingen University of Applied Sciences
Esslingen am Neckar, Baden-Württemberg,
Germany

ISSN 2191-530X ISSN 2191-5318 (electronic)
SpringerBriefs in Applied Sciences and Technology
ISBN 978-3-031-68750-1 ISBN 978-3-031-68751-8 (eBook)
https://doi.org/10.1007/978-3-031-68751-8

This Springer imprint is published by the registered company Springer Nature Switzerland AG
The registered company address is: Gewerbestrasse 11, 6330 Cham, Switzerland

If disposing of this product, please recycle the paper.

Preface

The book takes a deep dive into the industrial sphere, exploring subjects such as aerospace development, knowledge management in higher education, and the emergence of a nation as a player in the global space race. This insightful compilation of chapters offers an essential guide to navigating the complexities of modern industry, offering valuable insights and solutions to propel businesses and society toward a sustainable future.

Lumut, Malaysia   Azman Ismail
Machang, Malaysia   Fatin Nur Zulkipli
Lumut, Malaysia   Bakhtiar Ariff Baharudin
Esslingen am Neckar, Germany   Andreas Öchsner

Contents

Contents

Working Capital Management and Business Performance: A Study of Malaysian Small and Medium-Sized Family Enterprises

Nor Razuana Amram, Nurul Fadly Habidin, Noor Hidayah Zainudin, and Nazihah Wan Azman

Abstract Small and medium-sized enterprises (SMEs) in Perak, particularly family businesses, play a crucial role in contributing to the state's economy. Hence, this study examines the correlation between working capital management and business performance in Perak SME family businesses. A quantitative research approach was employed, and data were collected from 368 SME family businesses in Perak using simple random sampling techniques. Pearson correlation analyses were conducted using SPSS v.26. The results revealed a significant positive correlation ($r = 0.494$, $p < 0.05$) between working capital management and business performance, indicating that effective working capital management practices contribute to improved business performance in Perak SMEs family businesses. In conclusion, implementing working capital management in Perak SME family businesses is crucial for increasing business performance. This study recommends that businesses delve deeper into the dynamics of working capital management within Perak SME family businesses.

Keywords Working capital management · Performance · SME · Family business · SPSS

N. R. Amram (✉) · N. H. Zainudin · N. W. Azman
Faculty of Management & Information Technology, Universiti Sultan Azlan Shah, Kuala Kangsar, Perak, Malaysia
e-mail: norrazuana@usas.edu.my

N. H. Zainudin
e-mail: noorhidayah@usas.edu.my

N. W. Azman
e-mail: nazihah_azman@usas.edu.my

N. F. Habidin
Faculty of Management and Economics, Universiti Pendidikan Sultan Idris, Tanjong Malim, Perak, Malaysia
e-mail: fadly@fpe.upsi.edu.my

© The Author(s), under exclusive license to Springer Nature Switzerland AG 2024
A. Ismail et al. (eds.), *Technological Frontiers and Sustainable Innovations*,
SpringerBriefs in Applied Sciences and Technology,
https://doi.org/10.1007/978-3-031-68751-8_1

1 Introduction

Working capital management is a crucial aspect of financial management for small and medium-sized enterprises (SMEs) [1]. Effective working capital management can greatly impact the overall business performance and success of an SME [2]. Furthermore, proper working capital management can greatly impact SMEs' business performance and overall success, particularly family-owned ones [3]. The authors discovered that family businesses particularly face unique challenges and complexities regarding working capital management. These challenges may include balancing the competing interests of family members, ensuring efficient cash flow, and maintaining healthy relationships with suppliers and customers. Therefore, without efficient working capital management, family-owned SMEs may face significant difficulties sustaining their operations and achieving long-term growth [3].

Recently, Menteri Besar of Perak (Chief Minister of Perak), Datuk Seri Saarani Mohamad has launched Perak Sejahtera 2030 on June 22, 2022 [4]. Perak Sejahtera 2030 is a development plan aimed at promoting the overall well-being and prosperity of Perak, a state in Malaysia [5]. According to recent reports, Perak's gross domestic product (GDP) for 2021 is 3.5%, a positive sign for the state's economic development, indicating potential growth and investment opportunities [4]. Through Perak Sejahtera 2030, the government has established the Geran Usahawan Perak (GeRAK), allocating grants for all micro and small entrepreneurs in Perak whose annual sales do not exceed RM500,000 [6].

With the allocating of these funds, entrepreneurs in Perak will have the opportunity to invest in their businesses, expand their operations, and enhance their productivity. To ensure the efficient and fair distribution of the GeRAK grants, the Perak government has set up a transparent and accessible application process [6]. Entrepreneurs can submit their proposals outlining their business plans and how the grant will be utilized to improve their enterprises.

Furthermore, the Perak government has partnered with various financial institutions and business support organizations to provide additional resources and expertise to GeRAK recipients. This collaborative effort seeks to create a conducive ecosystem for entrepreneurship and economic development in Perak. With the implementation of GeRAK, the Perak government is committed to nurturing a vibrant and resilient entrepreneurial landscape, driving economic growth, and creating job opportunities within the state [6].

In line with this growth, Perak has positioned itself as one of the four top states attracting the majority of approved investments in the country [5]. The conducive business environment and strategic initiatives have undoubtedly contributed to this achievement. This presents an opportunity for businesses, particularly SMEs, to leverage the favorable investment climate and enhance their working capital management. In addition, by optimizing cash flow, inventory management, and receivables, SMEs can better position themselves to take advantage of the burgeoning economic prospects within Perak.

Furthermore, the identified problem is the lack of working capital management implementation in the day-to-day operations of Perak SME family businesses. Working capital is indispensable for financing the production cycle and essential capital expenditures necessary to maintain or expand current operations, playing a crucial role in the process [7]. Notably, good working capital management has been discovered to increase business performance [8]. However, a study by [9] reported poor working capital management among Malaysian SMEs leading to poor business performance. Therefore, systematically implementing working capital management is required to improve the business performance of Perak SME family businesses. Thus, this study aims to examine the correlation between working capital management and business performance in Perak SMEs' family business.

Presently, the resource-based view (RBV) theory explains how an organization's resources and capabilities contribute to its competitive advantage and long-term performance. According to the RBV theory, an organization's performance is determined by its resources and capabilities [10]. Similarly, [11, 12] proposed an explanatory RBV theory for working capital practices, and their study discovered a significant and positive impact of working capital practices on performance, supporting the assertions of the RBV theory.

RBV theory offers a compelling framework for understanding the relationship between working capital management and the business performance of SMEs [13, 14]. At its core, the RBV theory postulates that business can gain and sustain competitive advantage by strategically managing their unique bundle of resources and capabilities [13]. In the context of SMEs, working capital management represents a critical operational resource that can significantly affect a business overall performance and competitiveness.

Effective management of working capital—comprising cash, receivables, and payables—enables SMEs to maintain liquidity, reduce costs, and increase profitability [7]. By efficiently managing their cash flows, SMEs can ensure they have the required funds available for both daily operations and investment in increased opportunities. This is particularly crucial for SMEs, which often face more significant financial constraints and have limited access to external financing compared to larger corporations [15].

Efficient working capital management has been identified as a critical internal capability of a business that enhances its operational efficiency and profitability. Businesses that maintain key assets and liabilities optimally tend to have more stable operations leading to better performance. Therefore, RBV suggests that working capital management is vital to a firm's competitive advantage and improved performance [12]. Furthermore, the RBV theory can serve as a useful theoretical framework to understand the correlation between working capital management, access to finance, and SME performance [16].

The significance of working capital in the economy is evident through studies from various countries such as Malaysia [17], Argentina [15], Kosovo [9, 18], Sri Lanka [19], Bangladesh [20], Ghana [21], the UK [22], and Spain [23]. Working capital management involves the administration of current assets and current liabilities, crucial for ensuring the operational efficiency and financial health of a business [7].

For SMEs, particularly family-owned firms, effective working capital management is paramount due to their limited access to external funding and the need for preserving liquidity for uninterrupted operations [17].

Alternatively, [17, 19] established a positive correlation between working capital management practices and raised SME performance, highlighting the importance for SME owners to possess theoretical and practical knowledge. Government-led training programs could encourage effective working capital management among SMEs, enhancing their business performance [19]. Hence, it is essential for businesses to adopt working capital management practices to avail funding from both government and private entities.

There is increasing recognition within the literature of the crucial role of working capital management in the success and operational fluidity of businesses. For instance, within the agricultural SME sector, studies by [11] and [23] demonstrate the significant positive impact of working capital management on the performance of agricultural SME businesses. As [23] highlighted, incorporating effective working capital practices into business strategies is paramount for monitoring their impact on business performance.

Receivables are a crucial variable in working capital management, significantly impacting the profitability of SMEs [17]. Effective receivables management allows SMEs to enhance their cash flow, reduce the risk of bad debts, and improve overall financial performance by ensuring timely payment of receivables to meet financial obligations [7]. In addition, studies by [19, 20, 22] all reported that receivables management has a substantial impact on SME performance, with effective management leading to higher profitability. It is imperative for SME owners to focus on adopting effective receivables management, along with other working capital management practices, to improve their performance [19]. Therefore, closely monitoring accounts receivable and implementing effective credit policies are essential for SMEs to ensure timely customer collections.

To achieve effective receivables management, SMEs should implement a clear and transparent invoicing process to ensure prompt payment from customers [22]. This can involve setting clear payment terms, sending out invoices promptly, and following up on overdue payments. Additionally, offering incentives for early payment or implementing late fees for overdue payments can encourage customers to pay on time. Moreover, leveraging technology such as accounting software can streamline the receivables management process by automating invoicing, tracking payments, and sending reminders for overdue invoices [24]. This can help SMEs to monitor their outstanding receivables and take timely actions to recover any overdue payments.

Additionally, establishing strong relationships with customers can also facilitate timely payments. Open communication, building trust, and understanding the financial capabilities of customers can lead to mutually beneficial solutions for managing receivables effectively [2]. By adopting a proactive and organized approach to receivables management can significantly contribute to optimizing cash flow and overall financial stability for SMEs.

Furthermore, cash management emerges as a significant variable among the components of working capital management. Effective cash management enables

SMEs to handle their cash flows efficiently, which is crucial for their survival and growth. Previous studies have reported a positive relationship between cash management and SME performance [19, 25, 26] since holding cash and equivalent liquid assets provides SMEs with the flexibility required in their transactions.

Authors of [25] discovered a positive impact of cash holdings on the operational performance of businesses, highlighting the value of maintaining a cash buffer to handle unforeseen financial demands, especially during crises. This finding underscores the importance of liquidity management in businesses, as having a sufficient cash reserve can provide a crucial cushion during economic downturns or unexpected expenses. Additionally, maintaining adequate cash holdings can also signal financial strength to investors and creditors, potentially enhancing the company's overall financial stability [25, 27]. It would be valuable for businesses to further explore the optimal level of cash holdings based on their specific industry, risk profile, and growth opportunities to maximize the benefits identified [28].

Accounts payables also play a vital role in working capital management, representing the money a business owes its suppliers for goods or services purchased on credit. Studies by [18–20, 26] have discovered that credit days have a significant and direct impact on SME performance. Moreover, delaying payments for accounts payables can increase operational efficiency and profitability, effectively managing cash flow and ensuring sufficient cash for short-term obligations and long-term growth. Hence, effective management of accounts payables is critical for SMEs to optimize their cash flow and ensure financial stability [18].

As businesses continue to navigate through uncertain economic conditions, the management of accounts payables becomes increasingly important. One strategy that SMEs can employ to optimize their cash flow is to negotiate favorable payment terms with their suppliers [24]. By extending payment terms without incurring additional costs, businesses can improve their working capital and maintain healthy cash reserves. Additionally, implementing efficient invoicing and payment processes can help streamline the accounts payable function, reducing the risk of late payments and strengthening relationships with suppliers [2]. Moreover, leveraging technology and automation can further enhance the accuracy and efficiency of accounts payable management, allowing businesses to focus on strategic financial planning and growth initiatives.

The performance of SMEs family businesses can significantly benefit from adopting a RBV-centric approach toward managing working capital [17]. The authors found by recognizing and capitalizing on their unique familial and business relationships; these enterprises can negotiate better terms with suppliers, access preferable credit arrangements, and optimize their inventory levels to match demand more accurately. Such strategic management of working capital components not only improves liquidity and financial stability but also supports the business in seizing growth opportunities and enhancing overall competitiveness [29, 30].

Moreover, the intrinsic motivation and commitment often found in family members managing their enterprises can foster a culture of cost consciousness and operational efficiency [3]. This cultural aspect, viewed as a unique resource under RBV, can lead further to innovative working capital practices that are tailored to the specific needs and capabilities of the SME [11]. The alignment of such practices with the firm's overall strategy and resources can significantly contribute to its performance, enabling it to create a sustainable competitive edge in its market niche [31].

The working capital management in this study is focusing on three key constructs: accounts receivable, cash management, and accounts payable. These constructs are regarded as essential resources for businesses. The study aligns with the RBV theory, suggesting that Perak SMEs should adopt and implement these practices to enhance their business performance.

H_a1: Working capital management significantly correlates with the business performance of small and medium-sized enterprises (SMEs) family businesses in Perak.

2 Methodology

This study, a quantitative study with simple random sampling, was conducted, and the sample size of 368 family-owned SMEs was determined according to the method of Krejcie and Morgan [32]. Pearson correlation analyses were performed using SPSS v.26 to examine the correlation between working capital management and business performance in family-owned SMEs in Perak.

Perak was selected as the sample site due to the Perak Sejahtera 2030 development initiatives and the "1 Family, 1 Entrepreneur" program. The Prosperous Perak 2030 plan serves as a comprehensive government development strategy that aims to promote the potential of entrepreneurs as part of broader efforts to increase the capacity of the labor force and improve the socio-economic conditions and overall well-being of the people of Perak.

This study utilized business performance as the dependent variable, with working capital management as the independent variable. The construct of working capital management comprises accounts receivable, cash management, and accounts payable. Additionally, the questionnaires employed in this study were adapted from those used by [19].

This study concentrates on family owners or managers of registered businesses under the SME Corporation Malaysia that have been operating for over three years. The target population comprises 8,261 registered businesses under the SME Corporation as of December 31, 2021. This study was organized for a sample size of 368 but collected 450 responses to account for and remove outliers or anomalous data. After removing 70 outliers, the remaining 380 replies were more than enough to meet the needed sample size of 368. These businesses span various industries and are classified by state, as illustrated in Fig. 1.

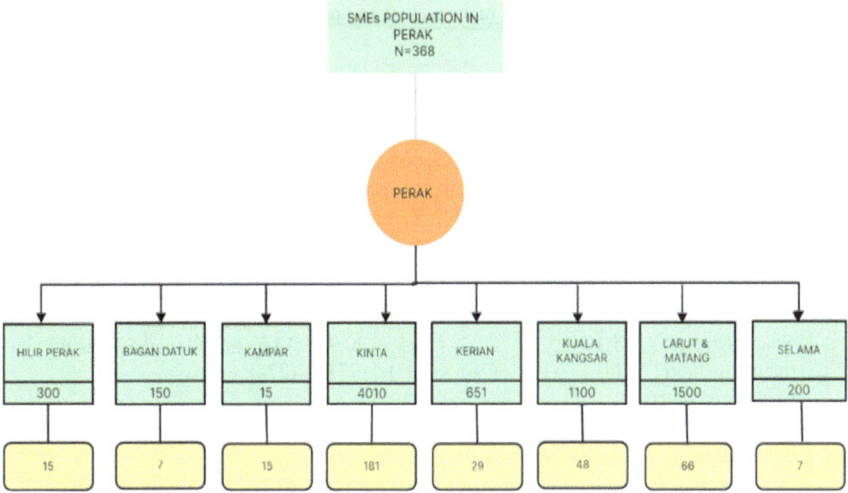

Fig. 1 Population sampling

3 Results

Table 1 presents a summary of the descriptive statistics for the analyzed variables. It outlines the distribution of working capital management and performance among SME family businesses. The findings indicate that 82.6% (304) of Perak SME family businesses, with a mean of 4.05, agree that the business evaluates the levels of receivables. Similarly, 78.0% (289) of Perak SME family businesses, with a mean of 4.00, agree on evaluating the levels of payables. Additionally, 73.9% (244) of Perak SME family businesses, with a mean of 3.78, agree on cash transactions align with the principles of the business cash budgets, while 73.5% (274), with a mean of 3.90, maintain liquid cash for its daily operational transactions. Furthermore, 69.0% (254) of Perak SME family businesses, with a mean of 3.70, prefer conducting payments to suppliers on a cash basis, and 67.7% (249), with a mean of 3.66, favor cash transactions over credit for customers. Lastly, 38.1% (140), with a mean of 2.86, agree on avoiding paying suppliers directly and instead opting for using a credit card, resulting in additional bank charges. On average, the working capital management and performance of family businesses stand at 68.7% (250), with a mean of 3.71.

Table 2 indicates the correlation between working capital management and business performance in Perak SMEs family business. The results reveal that there is a significant correlation between working capital management and business performance in Perak SME family business $r = 0.494^{**}$, sig 0.001 ($p < 0.05$). There exists a moderate correlation observed between the management of working capital and the performance of family-owned SMEs. This correlation is indicated to be positive through Pearson correlation analysis, suggesting that improved management of working capital corresponds to enhanced business performance. Acceptance of the

Table 1 Distribution of working capital management and performance of SME family businesses

No.	Statements	Scale					Total (368)	Mean	Sd	Level
		1 SDS	2 D	3 N	4 A	5 SA				
1	Your business ensures that cash transactions align with the principles of the business cash budgets	7.1 (26)	6.3 (23)	12.2 (45)	50.5 (186)	23.9 (88)	**73.9** **(244)**	**3.78**	1.10	High
2	Your business maintains liquid cash for its daily operational transactions	1.1 (4)	3.3 (12)	21.2 (78)	53.5 (197)	20.9 (77)	**73.5** **(274)**	**3.90**	0.80	High
3	Your business prefers cash transactions over credit for customers	4.1 (15)	17.7 (65)	10.6 (39)	43.5 (160)	24.2 (89)	**67.7** **(249)**	**3.66**	1.14	High
4	Your business evaluates the levels of receivables	0.8 (3)	6.0 (22)	10.6 (39)	52.4 (193)	30.2 (111)	**82.6** **(304)**	**4.05**	0.85	High
5	Your business conducts payments to suppliers on a cash basis	3.0 (11)	13.0 (48)	14.9 (55)	47.3 (174)	21.7 (80)	**69.0** **(254)**	**3.72**	1.04	High
6	Your business avoids paying suppliers directly and instead opts for using a credit card, resulting in additional bank charges	19.3 (71)	23.9 (88)	18.8 (69)	28.0 (103)	10.1 (37)	**38.1** **(140)**	**2.86**	1.29	Moderate
7	Your business evaluates the levels of payables	1.9 (7)	4.3 (16)	15.8 (58)	47.0 (173)	31.0 (114)	**78.0** **(287)**	**4.00**	0.90	High
	Total mean						**68.7** **(250)**	**3.71**	1.02	High

Strongly Disagree 2. Disagree 3. Neutral 4. Agree 5 Strongly Agree

Table 2 Correlation of working capital management and business performance in Perak SME family business

Variable	Pearson correlation	Sig (2-tailed)
Working capital management * business performance	0.494^{**}	0.001

alternative hypothesis, H_a1, occurs when a significant correlation between working capital management and business performance is established among Perak SMEs' family businesses. This acceptance is contingent upon obtaining a significance level (p-value) below 0.05. Conversely, rejection of H_a1 transpires when the significance level exceeds 0.05, indicating a lack of substantial correlation.

Working capital management plays a crucial role in the overall performance of Perak SME family businesses. The positive correlation coefficient of $r = 0.494^{**}$ indicates a strong relationship between effective working capital management and improved business performance. This suggests that efficient management of current assets and liabilities has a direct impact on the profitability and success of SME family businesses in Perak. Therefore, these businesses need to focus on optimizing their working capital to enhance their financial performance and sustainability in the long run.

4 Discussion

This study aims to examine the correlation of working capital management on the business performance in Perak SMEs family businesses. It seeks to confirm previous findings that indicate a significant correlation between working capital management and the business performance of SMEs [8, 11, 17, 19, 31].

Despite the lack of implementation of working capital management in day-to-day operations of family-owned SMEs in Perak, this study is motivated by Perak Sejahtera 2030. The government has established the GeRAK, allocating grants for all micro and small entrepreneurs in Perak whose annual sales do not exceed RM500,000 [6]. Therefore, this study has accepted the alternative hypothesis that working capital management significantly correlates with the business performance of Perak SMEs family business (H_a1).

This aligns with studies conducted by [19, 20, 22], which determine receivable management as a factor of working capital management, and the results demonstrate receivables management significant influencing performance of SMEs. Effective receivables management is crucial as it can lead to increasing performance of SME business. This study found that effective management of working capital plays a crucial role in the performance of family-owned small and medium-sized enterprises in the Perak region. The efficient handling of accounts receivable, accounts payable, and cash has a direct impact on the financial health and overall success of these businesses. Furthermore, this study highlighted the importance of implementing sound working capital management strategies to enhance profitability, liquidity, and operational efficiency. These findings emphasize the need for Perak SMEs to prioritize and optimize their working capital management in order to achieve sustainable growth and competitiveness in the market.

To achieve these outcomes, SME family businesses in Perak should prioritize developing and implementing comprehensive working capital management practices tailored to their specific industry, business model, and growth objectives. Thus, ensure the stability of their operations and drive continuous improvement in business performance, laying a solid foundation for long-term prosperity and resilience. The RBV theory offers valuable insights into how effective utilization of working capital can contribute to an SME's family business competitive advantage. By strategically deploying their current assets and liabilities, SMEs can leverage their internal resources to create sustainable competitive advantages, such as improved operational efficiency, enhanced cash reserves, and better supplier relationships.

Furthermore, the methodology and empirical aspects related to working capital management in the Perak SME family business allow researchers to delve into the specific challenges and opportunities in the region. Understanding the practical implications of working capital management for Perak SME family businesses, particularly those in the context of family businesses, can offer rich insights into the unique dynamics and how working capital management strategies can be tailored to suit the specific needs of family-owned SME family businesses in Perak. The results of this study are consistent with the RBV theory, and this study aligns with the previous studies by [11, 12], which employed the RBV theory, supporting this study's findings that working capital practices significantly and positively impact performance. Therefore, based on the findings of this study, working capital management serves as a source of business.

5 Conclusion

In conclusion, the results highlight the significant correlation between working capital management and business performance, indicating that managing working capital efficiently can lead to improved financial performance and overall success for Perak SMEs family businesses. This underscores the need for SMEs to prioritize and optimize their working capital management practices in order to achieve sustained growth and competitiveness in the market. Hence, integrating theoretical, methodological, empirical, and practical perspectives in understanding the contribution of working capital management to the performance of SMEs family businesses in Perak will enhance scholarly knowledge and provide actionable insights for businesses operating in the region.

The use of RBV theory in a theoretical perspective would involve delving into existing literature and theoretical frameworks related to working capital management and its impact on business performance, particularly within the context of family businesses in Perak. This can contribute to a deeper understanding of the underlying mechanisms and dynamics. In a methodological contribution, the analysis of actual data collected from SME family businesses in Perak allows for the validation and substantiation of theoretical propositions within the local context. This can contribute to the development of a more nuanced understanding of the relationship between working capital management and business performance in Perak SME family businesses.

Furthermore, an empirical contribution of the objective development can conduct statistical analyses and tests to determine the strength and direction of the relationship between working capital management and business performance. By examining financial indicators and performance metrics, the empirical perspective can provide concrete evidence of the impact of working capital management on the financial health and success of SMEs family businesses in Perak. Lastly, the practical perspective would involve synthesizing the theoretical, methodological, and empirical findings into actionable recommendations and strategies for enhancing working capital management practices within SME family businesses in Perak, ultimately leading to improved business performance.

Therefore, for further research, financial technology (fintech) could be crucial in improving access to capital for Perak SME family businesses. Fintech has emerged as a powerful tool that could significantly impact capital accessibility for Perak SME family businesses. Moreover, by leveraging fintech solutions, these businesses can gain access to alternative lending platforms, crowdfunding opportunities, and streamlined financial transaction processes. This, in turn, can lead to improved working capital management and business performance. In addition to traditional banking services, Perak SME family businesses can explore fintech options such as peer-to-peer lending, online invoice financing, and digital payment solutions. Accordingly, these technologies have the potential to provide quicker access to funding, lower transaction costs, and greater flexibility in managing cash flows.

Acknowledgements The funding for this paper is provided by a grant from the "Dana Khas Penyelidikan Universiti (DKPU) 2023" at Universiti Sultan Azlan Shah (USAS).

References

1. D.T.K. Bernard, A. Almeida, S. Perera, H. Jayarathna, A.A.S.N. Munasighe, The effect of working capital management on the export performance of small and medium export enterprises: evidence from export manufacturing sector in Sri Lanka. J. Econ. Bus. **2** (2019). https://doi.org/10.31014/aior.1992.02.03.117
2. A.K. Panda, S. Nanda, P. Panda, Working capital management, macroeconomic impacts, and firm profitability: evidence from Indian SMEs. Bus. Perspect. Res. **9**, 227853372092351 (2020). https://doi.org/10.1177/2278533720923513
3. M. Mokhber, G.T. Gi, S.Z. Abdul Rasid, A. Vakilbashi, Z.N. Mohd, S.Y. Woon, Succession planning and family business performance in SMEs. J. Manag. Dev. **36**, 330–347 (2017). https://doi.org/10.1108/jmd-12-2015-0171
4. Department of Statistic Malaysia: Perak (2023). https://v1.dosm.gov.my/v1/index.php?r=column/cone&menu_id=RTRycHhPcisweHpMdlVwKzhMY25XUT09. Accessed 13 Jan 2024

5. Malaysian Investment Development Authority: Perak in top four with RM7.1 billion approved investments in 2Q, says MB (2023). https://www.mida.gov.my/mida-news/perak-in-top-four-with-rm7-1b-approved-investments-in-2q-says-mb/. Accessed 12 Jan 2024
6. PUBI Perak: Permohonan Geran Usahawan Perak (GeRAK) (2023). https://www.pubiperak.com/permohonan-geran-usahawan-perak-gerak/. Accessed 13 Jan 2024
7. S.A. Ross, R. Westerfield, B.D. Jordan, *Fundamentals of Corporate Finance* (McGraw-Hill Education, New York, NY, 2022)
8. M.C. Hernandez, C.S. Balboa, R.C. Cuenca, N.D.G. Quilantang, Assessment of financial management practices of small and medium enterprises (SMEs) in Nasugbu, Batangas. Int. J. Creative Bus. Manag. (IJCBM). **1**, 24–37 (2021)
9. A. Ahmeti, Y. Ahmeti, S. Ahmeti, The impact of working capital management on SME profitability: evidence from Kosovo. J. Econ. Bus. **40**, 459–478 (2022). https://doi.org/10.18045/zbefri.2022.2.459
10. B. Wernerfelt, The resource-based view of the firm: ten years after. Strateg. Manag. J. **16**, 171–174 (1995). https://doi.org/10.1002/smj.4250160303
11. K.M. Mang'ana, D.W. Ndyetabula, S.J. Hokororo, Financial management practices and performance of agricultural small and medium enterprises in Tanzania. Soc. Sci. Humanit. Open **7**, 100494 (2023). https://doi.org/10.1016/j.ssaho.2023.100494
12. A. Bhattacharyya, M.L. Rahman, S. Wright, Improving small and medium-size enterprise performance: does working capital management enhance the effectiveness of financial inclusion? Account. Finance (2023). https://doi.org/10.1111/acfi.13081
13. R. Bhuyan, M.S.H. Khandoker, N. Tasneem, M. Taznin, Working capital management (WCM) and firm performance in emerging markets: a case of Bangladesh. Account. Finance Res. **10**, 36 (2021). https://doi.org/10.5430/afr.v10n1p36
14. H.C. Huan, D.N.M.H. Hoang, Working capital management and firm profitability: evidence from SMEs in Malaysia. J. Sci. Technol. Issue Inf. Commun. Technol., pp. 99–105 (2020). https://doi.org/10.31130/jst-ud2020-090e
15. L. Sensini, M. Vazquez, Effects of working capital management on SME profitability: evidence from an emerging economy. Int. J. Bus. Manag. **16**, 85 (2021). https://doi.org/10.5539/ijbm.v16n4p85
16. G.A. Afrifa, K. Padachi, Working capital level influence on SME profitability. J. Small Bus. Enterp. Dev. **23**, 44–63 (2016). https://doi.org/10.1108/jsbed-01-2014-0014
17. N.R. Amram, N.F. Habidin, M.F. Basri, The relationship between working capital management and business performance in Malaysia SMEs family business. Int. J. Acad. Res. Bus. Soc. Sci. **13** (2023). https://doi.org/10.6007/ijarbss/v13-i9/17910
18. A. Ahmeti, D. Balaj, Influence of working capital management on the SME's profitability: evidence from Kosovo. Calitatea **24**, 154–162 (2023). https://doi.org/10.47750/QAS/24.192.18
19. H.M.D.N. Somathilake, C. Pathirawasam, The effect of financial management practices on performance of SMEs in Sri Lanka. Int. J. Sci. Res. Manag. (IJSRM) **8**, 1789–1803 (2020). https://doi.org/10.18535/ijsrm/v8i05.em05
20. M. Wali, A. Zahid, I. Khan, N. Islam, Working capital management and SME profitability: empirical evidence from Bangladesh. Glob. J. Manag. Bus. **5**, 94–99 (2018)
21. J. Lamptey, A.B. Marsidi, B. Usman, A.B. Ali, W. Suleiman, Overconfidence bias in working capital management and performance of small and medium enterprises: the perspectives of Ghanaian SME managers. Int. J. Acad. Res. Bus. Soc. Sci. **10** (2020). https://doi.org/10.6007/ijarbss/v10-i6/7372
22. G.A. Afrifa, I. Tingbani, Working capital management, cash flow and SMEs' performance. Int. J. Bank., Account. Finance **9**, 19 (2018). https://doi.org/10.1504/ijbaaf.2018.089421
23. L. Rey-Ares, S. Fernández-López, D. Rodeiro-Pazos, Impact of working capital management on profitability for Spanish fish canning companies. Mar. Policy **130** (2021). https://doi.org/10.1016/j.marpol.2021.104583
24. M.M. Thottoli, A.H. Al-Shukaili, M. Ali-Alalawi, F.K. Al-Amri, Does creditors terms and accounting process affect msmes debtor's management? The need for novel IT tools. Rev. Gestão Inovação e Tecnologias **11**, 4545–4560 (2021). https://doi.org/10.47059/revistageintec.v11i4.2483

25. L. Rocca, E. Tiziana, SME cash holdings and performance in the EU energy industry: a moderating role of environmental performance, in *EURAM 2019 Exploring the Future of Management*, pp. 1–39 (2019)
26. T. Ramajeyam, L. Sooriyakumaran, T.L. Vannarajah, Financial reporting practices and performance and small medium enterprises (SMEs) in Jaffna District of Sri Lanka. Int. J. Res. Anal. Rev. (IJRAR) **10** (2023)
27. P.Y. Loo, W.T. Lau, Key components of working capital management: investment performance in Malaysia. Manag. Sci. Lett., 1955–1964 (2019). https://doi.org/10.5267/j.msl.2019.7.010
28. N.E.A. Mohamad, N.R.A. Rahman, N.M. Saad, Linking working capital policy towards financial performance of small medium enterprise (SME) in Malaysia. SHS Web Conf. **36**, 00021 (2017). https://doi.org/10.1051/shsconf/20173600021
29. N. Ahangar, Is the relationship between working capital management and firm profitability non-linear in Indian SMEs? Small Enterp. Res. **28**, 23–35 (2021). https://doi.org/10.1080/132 15906.2021.1872685
30. D.K. Chalmers, L. Sensini, A. Shan, Working capital management (WCM) and performance of SMEs: evidence from India. Int. J. Bus. Soc. Sci. **11** (2020). https://doi.org/10.30845/ijbss.v11n7a7
31. H. Tran, M. Abbott, C.J. Yap, How does working capital management affect the profitability of Vietnamese small- and medium-sized enterprises? J. Small Bus. Enterp. Dev. **24**, 2–11 (2017). https://doi.org/10.1108/jsbed-05-2016-0070
32. R.V. Krejcie, D.W. Morgan, Determining sample size for research activities. Educ. Psychol. Measur. **30**, 607–610 (1970). https://doi.org/10.1177/001316447003000308

Image Processing Recognition Technique for Quality Control (Conveyor System)

**Mohd Shahrizan Yusoff, Muhamad Sazali Said,
Anis Farzana Ahmad Fauzee, Mohamad Arif Alias,
Muhammad Danial Izazi Zulkiflee,
and Muhamad Izzul Akmal Mohd Zulkifli**

Abstract Quality control is the most critical step in the manufacturing industry to ensure that the product quality is up to par. Malaysia's industrial sector is moving towards industrial revolution 4.0, in which most machines are connected, and production processes are accelerated. Human resources are still used in most quality inspection sections in the manufacturing industry. This project aims to create a defect detection method for image processing using MATLAB and evaluate the algorithm's effectiveness. The techniques used is grayscale, image enhancement, and image segmentation. This device is able to detect a few types of defects on the product surface. Furthermore, this device will be at a lower cost than the current device. This project can be a solution to provide a powerful technique for the quality control process in industrial applications.

Keywords Defect detection · Image processing · Image segmentation · Quality control

M. S. Yusoff (✉) · M. S. Said · A. F. A. Fauzee · M. A. Alias · M. D. I. Zulkiflee ·
M. I. A. M. Zulkifli
Universiti Kuala Lumpur Malaysian Spanish Institute, Kulim Hi-Tech Park, 09000 Kulim, Kedah,
Malaysia
e-mail: mshahrizan@unikl.edu.my

M. S. Said
e-mail: msazali@unikl.edu.my

A. F. A. Fauzee
e-mail: anis.fauzee@s.unikl.edu.my

M. A. Alias
e-mail: mohamad.aalias@s.unikl.edu.my

M. D. I. Zulkiflee
e-mail: mdanial.izazi@s.unikl.edu.my

M. I. A. M. Zulkifli
e-mail: m.izzulakmal@s.unikl.edu.my

1 Introduction

Small mistakes can affect how well a product works in the manufacturing industry. One of the most common checks is to look for cracks, which are a common cause of failure. Finding mistakes during the manufacturing process is important for keeping the quality and accuracy of the product. It takes a long time and a lot of money to do an optical inspection. Scanning with a camera can be used to find problems automatically. People have come up with ways to process images because there are many odd shapes and sizes of cracks.

Some things are better when quality checks are done automatically instead of by hand. It is thought to be cheaper than hiring people to do the work. The computerised assessment can work nonstop for 24 h a day, seven days a week. The automated system can also get results that are very accurate. The whole process can be run better. These standards can be set with the help of mechanical quality management systems, and the application can watch computer vision cameras in real time. Automated inspection gets around the problems that come with systems that are checked by hand. One mistake that can be avoided is the wrong identification of a defect. These mistakes can also be traced back to the unreliability of human vision alone, the imprecision of human eyesight, and the cost of labour. Machine vision is better at measuring quality and amount than the human eye because it is faster and more accurate. Machine vision systems can find details on objects that are too small for humans to see and look at them more accurately. It also makes it possible for the system to do quality checks.

The problem statement was that quality inspections done by hand were not very accurate. Inspection work that is done by hand needs human judgement. The level of human judgement depends on how knowledgeable, prepared, and experienced the worker is [1]. This would also make it hard to keep the quality the same. Standardising quality is necessary to make sure that the product is perfect. Next, human error in manual quality inspection. Finding small flaws that are hard to see with the naked eye can lead to mistakes. A human inspection might find a problem that does not exist, and it is more likely to miss any real problems. Then, workers start to get tired. People's eyes tend to get tired, especially when they have to look at small things. Manual separation takes a lot of time and depends a lot on a person's ability to work nonstop for long periods of time. This can make your eyes and brain tired. Manufacturing workers who do the same tasks over and over again with the same movements for a long time may still make mistakes when the production manager tells them to do their jobs faster so that they can make more.

The main goal of this study is to look into what the defect detection technique is like. Image segmentation, enhancement, edge detection, and other things can be used to describe image processing. The next step is to make an algorithm that can find the flaws. MATLAB is the software that was used to make the algorithm. A camera is a tool that helps with image processing. Then, the algorithm for processing images is tested to see how well it works. The data is gathered and tested to see how well it works. A sample of the problem is tested to figure out what the result will be.

Using the model, a number of fake flaws will be made on the surface of the material. Defects are things like cuts, cracks, holes, and marks from folding.

Image segmentation is the method that is used. The RGB2GRAY function changed the image from a colour image to a grayscale image. Image segmentation is when you separate the foreground from the background or group pixels with similar colours or shapes together. Image thresholding is a quick and easy way to separate an image into its foreground and background. The function of graythresh used Otsu's method to generate a global threshold from a grayscale image. Otsu's method selects an entry that minimises the intraclass variance of the black-and-white pixels that have been thresholded. The show(I) function was used in this image processing output. It displays the grayscale image of the I. It employs the image data type's default display range and optimises the figure, axes, and image object characteristics for image display.

2 Literature Review

The visual inspection process is very important because it affects how well defects can be found. The process starts with looking for possible defects and putting them into groups. In the first stage, when a man looks at an item, he needs to be careful and have a more sensitive eye in order to spot possible mistakes. In the first stage of an inspection, the level of understanding and knowledge of the inspector is very important, as are reasonable working conditions and an awareness of possible flaws. The inspector has to decide if the product is good or bad based on what he or she knows about the type of flaw and how it is classified. At the end of the inspection process, the inspector decides if the development can move on to the next stage or if it needs to be set apart from other high-quality items [2].

People sometimes do not understand the situation well enough to make good decisions, process the data correctly, or do their jobs right. Over time, failure to perform will come from being tired and worn out. Fatigue can make it harder for people to concentrate and pay attention. Researchers [3] have found that headaches, eye pain, dry eyes, blurred vision, and back and shoulder pain are all signs of visual fatigue.

The environment or atmosphere of the workplace is very important for making sure that employees can do their jobs safely and comfortably. A worker's ability to do their job can be affected by things like loud noises, bad lighting, bad smells, changes in temperature, lack of cleanliness, and so on [4].

Image segmentation will make it easier to process images by telling the target from the background. It is a basic idea that makes image analysis and pattern recognition possible. There are different ways to divide an image into parts: the threshold method, edge detection, and region extraction. The threshold method is one of the most basic ways to separate parts of an image because it is easy to use and does not require much computing power. It is been used in a lot of different fields. The threshold methods are used to place target images in different grayscale ranges [5].

Fig. 1 Flow chart of the algorithm

3 Methodology

The image processing method will be done in MATLAB. Defect detection aims to develop an algorithm to detect a defect on the surface as shown in Fig. 1.

The camera on the Raspberry Pi was used to get the picture. The picture was taken by the camera, and it was sent to MATLAB files. With a good camera, you can get high-resolution images, which can help with wrong image analysis. The RGB2GRAY function changed the idea from a coloured picture to a grayscale picture. In this step, the RGB2GRAY function was put to use. The process reads and shows an RGB image and then turns it into a grayscale image. It keeps the image's brightness but takes away its colour and saturation.

The Otsu's method was used by the graythresh function to make a global threshold from a grayscale image. Otsu's method chooses a point that minimises the difference between the black-and-white thresholded pixels.

Different kinds of flaws were made in the samples before they were used. The piece of weakness was chosen based on the defect that happened most often. Using the sample, a number of fake flaws will be made on the surface of the material. Defects are things like cuts, cracks, holes, and marks from folding. The defects were then put through a test run on the conveyor using the algorithm.

The samples that were made were tested, and the results were written down. Using an algorithm for image processing, different kinds of flaws were checked for. The data collected were used to figure out how well it worked. A total of 28 samples were used to collect the data. Five different kinds of flaws were found in the samples. The list of flaws is in Table 1.

After all the data has been collected, it will be looked at. The data will be evaluated based on what was found and what was not. For the algorithm's efficiency, the number of successes and failures was used to figure out how well it worked.

4 Result and Discussion

The algorithm was made by first reading the image that was given as input. rgb2gray was used to change the colour image into a grayscale image, which was then used to process the image. Then, the image enhancement and segmentation functions were put to use. Last, the results of finding the flaws were shown.

To make the pictures stand out more, the picture needs to be changed to grayscale. It goes from a coloured image to a grayscale image by getting rid of the hue and

Table 1 Type of defect

Type of defect	Image	Type of defect	Image
Cut		Hole	
Scratch		Folding mark	

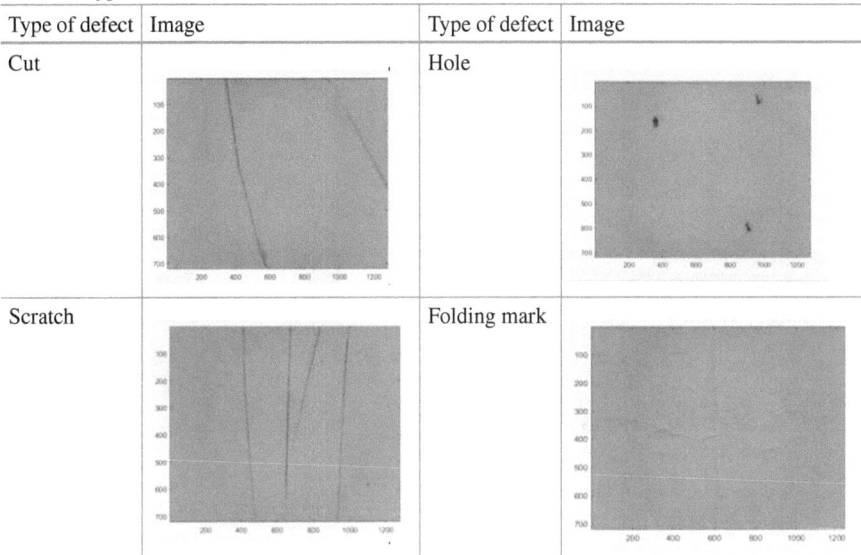

saturation information but keeping the brightness. This grayscale method was the best way to process pictures. Image segmentation was easy to do with grayscale images.

Then, the Otsu method was used to use the threshold function. It chooses the threshold to reduce the difference between black-and-white pixels in the same class. Based on the entry, the picture was turned into a binary picture. The function BW = im2bw turns a grayscale image into a black-and-white one. It does this by giving the value 1 (white) to pixels in the input image that have a brightness level greater than 0. Other pixels get the value 0 (black) (black). Figures 2 and 3 show how the images were changed.

The red line around the defect shows the end result of the defect detection. There are no red lines on the standard images where there are no problems. For the other mistake, it made red stripes around the mistake. There were some mistakes that were not marked correctly. The scratch defect and the hole defect were the only ones that could be found.

The image processing algorithm was able to find flaws on 2D surfaces in some way. There were some flaws that were clear. When the picture was taken in a bright place, it was easy to see the right mistakes. The light source should be bright enough to reach all of the sample's surfaces. Scratches and holes are the easiest flaws to find. Most of the scratches and holes were in places where they were easy to spot. The scuffs and holes also stood out against the light.

When they were taken in a low-light environment, there were flaws that could not be seen. When a picture is taken in a low-light area, the image will be dark and be seen as a flaw because the shadow is dark. It was hard to see the flaws in the folding

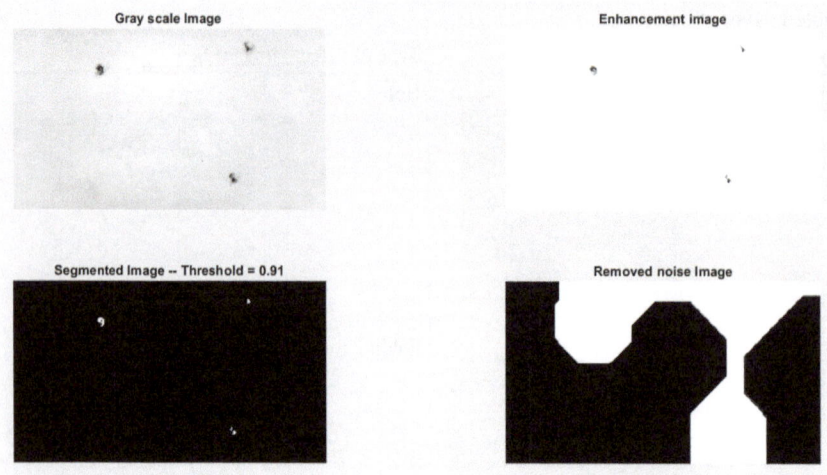

Fig. 2 Image processing of the defect

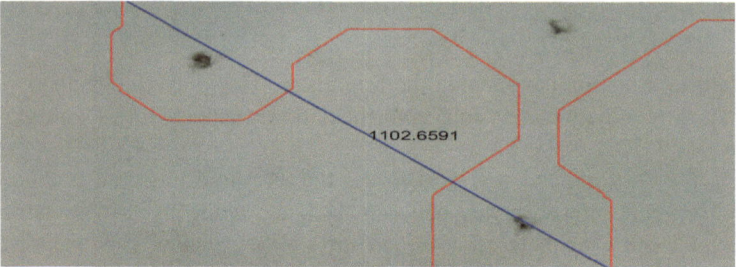

Fig. 3 Result of the defect detection

because the lines were barely visible on the surface. The watermark flaws, on the other hand, only had a stain of water on the surface and did not show the line.

Images that are too blurry can also make images that can not be seen. So that the image processing can find the flaw easily, the idea of the flaw needs to be as clear as possible.

Using the device and the algorithm, there have been 28 samples found. The results of the detection will be collected, and the data in Tables 2 and 3 will be used to do an analysis. Some examples are a cut, a hole, a scratch, a mark from folding, and a mix of these. The defects that were not found were in 5 samples, and the ones that were found were in 23 samples. The error rate for problems that were not found was 17.86%, and the error rate for problems that were found was 82.14.

The folding mark and watermark were the flaws that could not be found. The spot where it folded was hidden because the pattern of folding was hard to see on the camera. The light colour of the defect also made it hard to see the watermark on

Table 2 Error of detection

Type of defect	Number of detection	Percentage of correctness (%)
Not detected	5	17.86
Detected	23	82.14

Table 3 Type of undetected defect

Type of undetected defect	Image
Folding mark	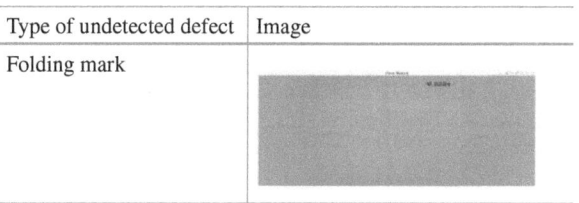

the camera. It is hard to get the right segments when there is not much difference or overlap between the grayscale values of the pixels.

For MATLAB to work with the camera system, it needs to be connected to the Raspberry Pi. The first step is to get the Raspberry Pi and MATLAB to work together. The Raspberry Pi needs to have all the latest updates, and the MATLAB file package needs to be put in place. As part of the algorithm, the resolution, mode, and time, it takes to take a picture were set.

After the Raspberry Pi Camera took a picture and MATLAB was used to process the picture, the segregation system started to work. If MATLAB finds a mistake, the segregation system will be moved to the left side of the conveyor. If MATLAB does not notice the problem, the conveyor will keep moving. When the image processing saw a problem with the product, the servo pushed it with the help of the divider. The algorithm must be in sync with the algorithm for processing images, so that the servo will run after the image processing algorithm has finished. For the segregation to work, the defect that was found must be bigger than 5 mm. If the impact of a found flaw is less than 5 mm, it would not be separated.

5 Conclusion

This project successfully developed an algorithm for image processing to detect a defect on the surface. Using an image processing technology enhances the images and segments the fault in the photos. The weaknesses of cut, scratch, and holes are mainly detected. The algorithm's effectiveness for image processing was high, as much as 82.14%. This technology will aid in the reduction of human mistakes in quality checking. It also aids in the uniformity of quality control inspections.

Acknowledgements A grant from Universiti Kuala Lumpur provided financial support for this study.

References

1. E.D. Megaw, Factors affecting visual inspection accuracy. Appl. Ergon. **10**, 27–32 (1979)
2. A. Kujawińska, K. Vogt, Human factors in visual quality control. Manag. Prod. Eng. Rev. **6**(2), 25–31 (2015)
3. H. Rajabi-Vardanjani, E. Habibi, S. Pourabdian, Designing and validation a visual fatigue questionnaire for video display terminals operators. Int. J. Prev. Med. **5**(7), 841–848 (2014)
4. Y. Ngadiman, B. Hussin, A.T. Bon, H.N.A. Abdul, Factors that influenced the quality inspection on the production line in manufacturing industry. MATEC Web Conf. **95**, 4–7 (2017)
5. Y. Zhang, The design of glass crack detection system based on image preprocessing technology. in *2014 IEEE 7th Joint International Information Technology and Artificial Intelligence Conference ITAIC*, pp. 39–42 (2014)

Case Study of a Monitoring Energy Consumption Device at a Residential Area

Adnan Bakri, Mohd Khairi Lutfi Shukri, Mohd Zul-Waqar Mohd Tohid, Mohd Al-Fatihi Sajudi, Munir Al-Faraj Al Akbir, Mohamad Shahrul Effendy Kosnan, Mohd Anuar Ismail, Zulhaimi Mohamad, Rahimah Kassim, Ahmad Nur Aizat Ahmad, and Izatul Husna Zakaria

Abstract In these most recent years, energy consumption has emerged as the most important concern. Consuming more electricity is a contributor to the rising energy demand, which is one of the variables impacting the increase in energy consumption. According to the statistics, one of the most significant contributors to overall energy consumption is the residential sector. In addition, the widespread outbreak of the Coronavirus disease (COVID-19), which is occurring right now all over the world,

A. Bakri (✉) · M. K. L. Shukri · M. Z.-W. M. Tohid · M. A.-F. Sajudi · M. A.-F. Al Akbir · M. S. E. Kosnan · M. A. Ismail · Z. Mohamad
Plant Engineering Technology Section, Universiti Kuala Lumpur Malaysian Institute of Industrial Technology, Persiaran Sinaran Ilmu, Bandar Seri Alam, 81750 Masai, Johor, Malaysia
e-mail: adnanb@unikl.edu.my

M. K. L. Shukri
e-mail: adnan.hjbakri@gmail.com

M. Z.-W. M. Tohid
e-mail: mzulwaqar@unikl.edu.my

M. A.-F. Sajudi
e-mail: mohdalfatihi@unikl.edu.my

M. A.-F. Al Akbir
e-mail: munir@unikl.edu.my

M. S. E. Kosnan
e-mail: mshahruleffendy@unikl.edu.my

M. A. Ismail
e-mail: manuar@unikl.edu.my

Z. Mohamad
e-mail: zulhaimi@unikl.edu.my

R. Kassim
Technical Foundation Section, Universiti Kuala Lumpur Malaysian Institute of Industrial Technology, Persiaran Sinaran Ilmu, Bandar Seri Alam, 81750 Masai, Johor, Malaysia
e-mail: rahimahk@unikl.edu.my

© The Author(s), under exclusive license to Springer Nature Switzerland AG 2024   23
A. Ismail et al. (eds.), *Technological Frontiers and Sustainable Innovations*,
SpringerBriefs in Applied Sciences and Technology,
https://doi.org/10.1007/978-3-031-68751-8_3

is having a significant influence on the energy industry. Energy is one of the most important factors in Malaysia's efforts to achieve its growth and development objectives in a sustainable manner. Through the years, substantial strategic planning has gone into ensuring the long-term viability of energy resources, and energy policies have been devised following an in-depth analysis of both the supply of energy and the demand for it in the present and the future, by making sure that the resources, especially the energy, are used effectively. We can save energy, which will lead to a decrease in the output of energy and, as a result, a reduction in the amount of carbon dioxide emissions that are produced. To determine the need for an energy monitoring device to be implemented in residential areas, the researcher has held an interview session and distributed questionnaires. The purpose of this study on energy monitoring devices in residential areas is to provide an indicator and raise consumer awareness regarding energy consumption.

Keywords Energy consumption · Energy efficiency · Residential area · Monitoring device

1 Introduction

Electricity is a necessity to modern civilization. It is a powerful source that has a significant impact on human daily activities [1–3]. Nowadays, there is an increasing demand for electricity energy worldwide. Such trend is proportionated with the growing of human population [4]. Of significant, the spike in the number of residential areas in Malaysia has a massive impact on the country's economic growth; ironically, it also raises the electrical energy demand [4]. The purpose of study is to determine the need for a monitoring device to regulate and manage the electrical energy consumption in residential areas.

A. N. A. Ahmad
Faculty of Technology Management and Business, Universiti Tun Hussein Onn, Malaysia, 86400 Parit Raja, Batu Pahat, Johor, Malaysia
e-mail: aizat@uthm.edu.my

I. H. Zakaria
School of Technology and Logistics Management, Universiti Utara Malaysia, 06010 Sintok, Kedah, Malaysia
e-mail: izatul.husna.zakaria@uum.edu.my

2 Literature Review

The energy consumption. The electricity consumption in Malaysia has increased rapidly over the last few decades, due to increasing in the number of residential areas.

Service and residential sectors are experiencing the greatest increase in term electricity and utility bills. The electricity industry accounts for 43.6% of all consumers, with natural gas accounting for the rest. Malaysia has an abundance of natural resources, such as natural gas. Most gas power stations are linked by pipeline gas.

Energy demand. Malaysia's energy consumption is predicted to reach 116 million tons of oil equivalents (MTOE) by 2020 [5]. The government's policy adjustment raised Malaysia's energy consumption. From an agricultural-based to a technological-based economy, Malaysia's growing population may lead to a rise in energy consumption. Malaysia saw a reduction in peak electrical demand because to COVID-19 last year, compared to the demand projected prior to the epidemic. When the country was afflicted by the epidemic, the peak demand for power in the peninsular Malaysia in 2020 was 18,808 MW. The overall power consumption for 2025 is predicted to be 19,365 MW, with a net demand of 18,442 MW in the same year. Malaysia's post-COVID-19 economic recovery will have an impact on the country's peak power consumption, which is expected to approach pre-COVID-19 levels by 2025 [4].

Energy consumption at residential area. Furthermore, Sena et al. [6] claimed that the influence of techno-socioeconomic factors also has an impact on the growing energy consumption in the residential sector. The sociodemographic, home characteristics, occupant behavior in purchasing and utilizing appliances, and appliance characteristics comprise the techno-socioeconomic. The number of inhabitants, family composition, age, level of education, household income, and work status all appear to be socio-demographic determinants influencing the residential energy usage. The number of residents influences power usage; therefore, the more people who live in a residence, the more electrical energy is consumed [3]. The effects and challenges of COVID-19 pandemics on energy demand and consumption highlight consumer-related energy lessons. The International Energy Agency (IEA) predicts that the 2020 energy demand shock will be the greatest in the previous 70 years. According to Ali et al. [1], the global energy demand is expected to fall by 6% in 2020 compared to 2019. As individuals are forced to stay at home, household energy usage rises. This is due to the fact that many governments only allow critical businesses to operate.

3 Methodology

Research methodology: This is a collection of methods for conducting systematic research. It simply means a guide to research and how to operate it. Before conducting research, the researcher must first identify the problems that arise at the target location. The researcher must also consider the research problem in the case studied. Other than that, the researcher must focus on the overall causes of problems and relate the problem to the related theory as well as evidence from the case study that combines theory and observation. In this research, the method used may include research on interviews, surveys, journals, and other research methods.

Target respondent. The researcher needs to meet with the respondent in order to obtain the data and information related to research field. Other than that, the researcher also targeted 50 respondents to gather the information based on questionnaire. Before conducting the interview and distributing the questionnaire, the researcher needs to follow the procedures. Figure 1 shows the procedure that the researcher needs to follow.

Analysis of the data from the interview. Figure 2 illustrates an overview of the step-by-step data analysis process, which was determined by focusing on the most significant aspect of the responses.

Data from questionnaire. The data collected during the survey process will be conducted using a Google form to survey regarding the importance of an energy monitoring device at residential area. After the target of respondents is achieved, the data is then transferred into Statistical Packages for Social Sciences (SPSS). By using this method, all the gathered data from the respondents will be presented descriptively.

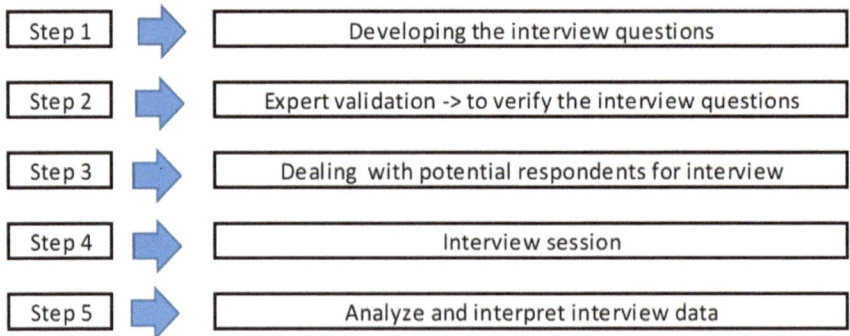

Step 1	➡	Developing the interview questions
Step 2	➡	Expert validation -> to verify the interview questions
Step 3	➡	Dealing with potential respondents for interview
Step 4	➡	Interview session
Step 5	➡	Analyze and interpret interview data

Fig. 1 Interview and questionnaire procedure

Fig. 2 Data analysis step in qualitative method

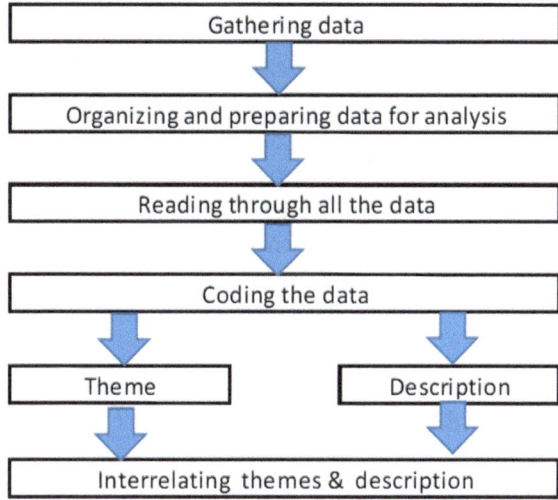

4 Data Analysis

Qualitative approach. The data was gathered using the qualitative method of interviewing and questionnaire. The data and information gathered during the interview session were converted from audio to sentence. The researcher will then analyze the data and information using a coding process based on the appropriate themes and formula. For the questionnaire, the researcher will use SPSS to analyze the data from surveys. The researcher has used an interview as a method to collect the information from the respondent. For the interview session, the researcher has interviewed two respondents from different backgrounds. Respondent 1 is working at Alliance Bank Malaysia Berhad as senior executive and staying in the Teratai Mewah Condominium at Setapak, Kuala Lumpur. For respondent 2, he is working as data analyst at Maxis, Petaling Jaya, and staying at Icon Residenz at Setapak, Kuala Lumpur.

Trouble on managing electricity bill at residential areas. There are 62% of respondents accounting to 31 respondents have problem on managing the energy consumption at their residential, followed by 22% accounting to 11 respondents did not have trouble on handling energy consumption at their residential. The rest of respondent which is 16% that accounting to 8 respondents said they may not have any trouble on managing energy consumption at residential. Therefore, most of the respondents from the survey are having trouble on managing energy consumption at residential.

Main priority for energy monitoring at residential areas. According to the interview, 38.8% of the respondents said they 'strongly agree,' followed by 34.7% of the respondents said they 'agreed.' The rest of which are 22.4, 2 and 2% of the respondents choose 'neutral,' 'strongly disagree,' and 'disagree,' respectively. Overall, the majority of the respondents state that they strongly agree that the device of energy monitoring becomes priority used to monitor energy consumption.

Opportunities of installing the energy monitoring device. Approximately, 48% of the respondents said they 'agree,' followed by 36% of respondents said they 'strongly agree.' The rest of which is 16% from the respondents choose 'neutral,' respectively. In short, the majority of the respondents state that they agree that the real-time monitoring can lead to lower the energy conservation by using energy monitoring device in residential areas. Based on the interview, 44% of the respondents said they 'often,' followed by 38% of respondents said they 'rarely.' The rest of which are 12, and 6% of the respondent choose 'very often' and 'never.' In short, the majority of the respondents state that they were often facing problems in managing energy consumption at residential areas.

5   Results and Discussion

To assure the successful conclusion of this investigation, three research questions (RQ) and research objectives (RO) were identified and built to address the viewed as an indication. The research questions and objectives were linked to the appropriate research approach so that the information could be obtained, and the research objectives could be satisfied. As a direct consequence of this, the outcomes and findings of each RQ and RO were discussed in more depth.

6   Limitations and Recommendation for Future Research

The most significant barrier is time, as the researcher has only 17 weeks to do this research. In addition to this research, the researcher is responsible for completing the following, another assignment, project, classes, and other tasks. This will influence this research because the researchers will need to allocate their time and focus their attention appropriately. Other than that, the limitation for researchers to set the schedule of an interview session with the relevant respondent. In the future, this research can be upgraded using even more advanced approaches. The research of monitoring energy consumption also can be applied to other sectors such as industrial and office areas. The research also can be able to penetrate heavy industry market that consumes higher-rated energy to meet the characteristic of energy efficiency.

7   Conclusion

In a nutshell, the primary objective of the research was to study the energy monitoring device, as well as to raise the level of awareness among residents regarding the significance of energy consumption and to assist in the changing of their attitudes and behaviors. As a consequence of this, users will control their electrical appliances in effective ways, which will lead to a reduction in the amount of electricity that is consumed as well as a decrease in the cost of the monthly bill.

References

1. M.A. Wahab, N.A. Ramli, Lighting control system for energy management system and energy efficiency analysis. J. Phys. Conf. Ser. **1529**(5), 52000–52022 (2020)
2. W.S.W. Abdullah et al., The potential and status of renewable energy development in Malaysia. Energies **12**(12), 2437 (2019)
3. M.S. Ahmed et al., Awareness on energy management in residential buildings: a case study in Kajang and Putrajaya. Eng. Sci. Technol. **12**(5), 1280–1294 (2017)
4. S.B.M. Ali et al., Analysis of energy consumption and potential energy savings of an institutional building in Malaysia. Alex. Eng. J. **60**(1), 805–820 (2021)
5. M. Hasanuzzaman et al., Global electricity demand, generation, grid system, and renewable energy polices: a review. Wiley Interdiscip. Rev **6**(3), 805–820 (2017)
6. B. Sena et al., Conceptual framework of modelling for Malaysian household electrical energy consumption using artificial neural network based on techno-socio economic approach. Int. J. Electr. Comput. Eng. **8**(3), 2088–8708 (2018)

An Efficient Fuzzy c-Means Neural Network Approach for the Prediction of Student Cheating Tendency for Online Learning System

Siti Fairuz Nurr Sadikan, Mohd Aliff Afira Sani, Sulaiman Mahzan, Mohd Ab Malek Md Shah, Azizul Azhar Ramli, and Mohd Farhan Md Fuzee

Abstract Due to its adaptability and versatility, online education has gained widespread popularity among university students. Students can customize their education to fit their schedules and lifestyles with the help of this technology. It is worth noting, however, that even the brightest student may feel pressured into cheating to earn a perfect score. If a student is having trouble finishing an assessment before the due date, or if the assignment is uninteresting, irrelevant, or very difficult, they may resort to cheating to get it done. The main objective of this study is to recommend the use of a fuzzy c-means neural network for forecasting student online cheating behavior. The students' identities were trained and verified in this manner to make predictions about their propensity to cheat. Fuzzy c-means neural networks were found to have a high degree of accuracy when applied to these datasets. Thus,

S. F. N. Sadikan · S. Mahzan · M. A. M. M. Shah
Universiti Teknologi MARA, Cawangan Melaka Kampus Jasin, 77300 Merlimau, Melaka, Malaysia
e-mail: fairuznurr@uitm.edu.my

S. Mahzan
e-mail: sulaiman@uitm.edu.my

M. A. M. M. Shah
e-mail: malek625@uitm.edu.my

M. A. A. Sani (✉)
Quality Engineering Research Cluster, Universiti Kuala Lumpur Malaysian Institute of Industrial Technology, 81750 Masai, Johor, Malaysia
e-mail: mohdaliff@unikl.edu.my

A. A. Ramli · M. F. M. Fuzee
Faculty of Computer Science and Information Technology, Universiti Tun Hussein Onn Malaysia, 86400 Parit Raja, Batu Pahat, Johor, Malaysia
e-mail: azizulr@uthm.edu.my

M. F. M. Fuzee
e-mail: farhan@uthm.edu.my

© The Author(s), under exclusive license to Springer Nature Switzerland AG 2024   31
A. Ismail et al. (eds.), *Technological Frontiers and Sustainable Innovations*,
SpringerBriefs in Applied Sciences and Technology,
https://doi.org/10.1007/978-3-031-68751-8_4

this method can aid educators like lecturers and teachers in assessing and grading each student's test and assignment and lowering the prevalence of cheating. The methods also represent a novel approach to ensuring the safety of distance education.

Keywords Online learning · Cheating · Fuzzy c-mean · Neural network

1 Introduction

The use of hybrid models that combine online and in-person training appears to be on the rise [1]. This technology can be used to tailor a student's online learning experience to their individual needs at any moment.

In the recent two decades, e-assessment, or online learning assessment, has become increasingly common in higher education [2]. As a result, schools can keep tabs on their pupils' development and performance [3]. It provides an alternative means of assessment for students who, for various reasons, are unable to attend regular classes.

Pressures to perform well or to escape the shame of failure might encourage even the most morally upstanding child to take the easy way out and resort to cheating. Many students, depending on the circumstances, cheat by very little amounts. Some students may turn to cheating on an exam if they are pressed for time, the material is uninteresting, or the task is very complex. Exams are the universal benchmark form of assessment in all higher education systems across the globe, making them particularly vulnerable to cheating [4]. However, every year there is an increase in the number of students who cheat on their exams.

Furthermore, the previous study found that gender (males have a higher tendency compared to females), age, study mode (part-time students have a higher tendency compared to full-time students), study level (diploma, degree, master's degree, doctorate), year of study (first year, second year, etc.), residency status (college stay or drop), grade point average (GPA), and pressure are the main contributing factors toward student cheating tendency [5–7].

Using an efficient fuzzy c-means neural network approach, this study aims to create a system for predicting student cheating based on the integrated authentication framework as in [8]. The purpose of this research is to better secure online education by utilizing student identity. For the purposes of this study, student identities will be used to monitor predicted instances of cheating. Students' gender, academic year, cumulative GPA, perceived academic pressure, and time spent on research are the only demographic variables considered in this study.

2 Literature Review

Artificial neural networks (ANNs) have lately found use in a broad variety of contexts due to their capacity to discern complex noisy patterns. Neural networks have their own limitations when used in place of tried-and-true methods like statistical regression, pattern recognition, and time series analysis. Before a neural network to accurately categorize new data, it needs a massive amount of training data. This constraint prevents the neural network from effectively training, which in turn reduces the detection accuracy [9].

Many different hybrid approaches have been investigated in the past; one such hybrid utilizes a neural network and fuzzy logic to compensate for each method's weaknesses. When it comes to classifying new data, neural networks excel, but fuzzy clustering helps the algorithm to generalize well [10]. The efficiency of the system is enhanced by integrating a neural network with fuzzy logic.

Lung cancer is hard to find early because most cancer cells are made up of structures that cover each other. There is a study that suggests that the Hopfield neural network (HNN) and a fuzzy c-mean (FCM) grouping algorithm can be used to find early lung cancer using color images of sputum [11]. Using the HNN and FCM methods, images with N pixels can be put into M different groups. They tested how well HNN and FCM did at classifying a set of 1000 color images of phlegm and found that HNN did a much better job.

In addition, Jithish has built an adaptive neuro-fuzzy inference system (ANFIS)-based hybrid intelligent system for estimating household water consumption [12]. The suggested system is supervised and trained to model how various environmental conditions affect residential water usage. Using data on air pressure, temperature, relative humidity, and wind speed, the system estimates how much water will be needed for household use in the future. The system's prediction of household water use is more accurate, as shown by an evaluation on a real-world smart home dataset.

The term "web mining" refers to the process of finding and analyzing relevant data on the internet. Web mining can be broken down into two main components, web contents mining and web usage mining, based on the distinct foci and methods used to acquire information. The term "web content mining" refers to the practice of using search engines and web spiders to automatically scan and retrieve data and content from millions of websites and online databases. Lakheyan and Kaur had reviewed online usage mining and its features and find that the fuzzy c-means technique is the most effective at retrieving data for search engines [13, 14].

3 Methodology

As can be seen in Fig. 1, the framework consists mostly of data preprocessing for training and testing data, a fuzzy c-means neural network, and a result for predicting students' propensity to cheat.

Fig. 1 Sample 1 (example)

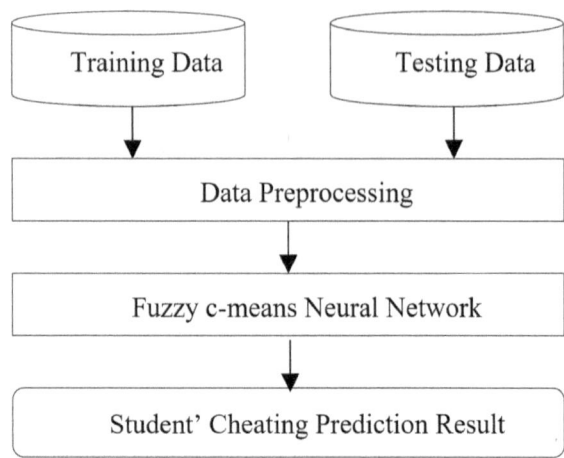

Data from a survey administered during the fall 2018–winter 2019 academic semester (September 2018–January 2019) was used to conduct experimental evaluations of the presented framework for cheating tendencies. The study's dataset was collected using a survey. UiTM Cawangan Melaka, which encompasses Jasin, Bandaraya, and Alor Gajah, was the site of the research.

The user information has been gathered with the help of a questionnaire that consists of ten questions. This tool consists of two sections: Part A (student demographic profiles) and Part B (influencers on students' propensity to cheat). The information was gathered from the samples using a form handed out to them during their visits to educational institutions. During data analysis, MATLAB, version R2017b, was used to evaluate the survey data.

4   Results and Discussion

Here, we see how a neural network method and fuzzy c-mean can shed light on users' dishonest tendencies. The same dataset is used, with the same number of inputs, including gender, year in school, GPA, pressure to achieve good grades, study time for any class, and hours spent.

The result is a yes/no indicator of cheating propensity. If status is positive, then dishonest behavior is present; otherwise, it is not. Click the FCM + NN tab in the left-hand navigation to view the FCM + NN analysis page.

An example of FCM + NN training on user cheating tendencies is displayed in Fig. 2 for the 80:20 dataset. The FCM parameter training iterations for this sample are set at 10. Meanwhile, for the sake of NN training, the starting weights and biases are chosen at random. The retrieved features and ten hidden layers are then used to create user models, with binary sigmoid serving as the activation function. The goal is an error of 10-6 over a period of 10 epochs, with a learning rate of 0.1.

training_fcm_nn —

Home Page Testing FCM+NN

FCM Parameters

Exponent — 2

Iterations — 10

Improvement — 1e-5

Weight

● Random ○ Load Weight

Processing

Load Data	Clustering
Load Weight	Training
Save Weight	Save Net+Center

Hidden Layer

Num. of Neurons — 10

Activation Func. — Binary Sigmoid

Output Layer

Num. of Neurons — 1

Activation Func. — Linear

Parameters

Training Function — trainlm

Error Goal — 1e-6

Epochs — 10

Learning Rate — 0.1

	Gender	Year	CGPA	Grade	Class	Prepare	TC (Actual)	TC (Predicted)
1	0	3	0	1	0	3	1	1
2	1	3	0	1	0	3	1	1
3	0	2	0	0	0	3	0	0
4	0	2	0	1	0	3	0	0
5	0	2	0	0	0	2	0	0
6	0	2	0	1	0	2	0	0
7	1	2	0	1	1	3	0	0
8	0	1	0	1	1	2	1	1
9	0	1	0	1	0	2	1	1
10	0	3	0	1	0	3	1	1
11	0	1	0	1	0	4	1	1
12	1	1	0	0	0	1	1	1
13	1	1	0	1	1	3	1	1
14	0	1	0	1	1	2	1	1
15	0	1	0	1	1	3	1	1
16	0	1	0	0	1	3	1	1
17	1	2	0	1	1	5	0	0
18	1	2	0	1	1	3	0	0
19	0	1	0	1	1	3	1	1
20	0	1	0	1	0	5	1	1
21	1	3	0	0	0	3	1	1
22	1	1	0	1	0	2	1	1
23	0	1	0	0	0	4	1	1
24	1	1	0	1	0	4	1	1
25	1	1	0	1	0	2	1	1
26	1	2	0	1	1	3	0	0
27	0	1	0	1	1	3	1	1

Accuracy

Num. of Data 1128 Num. of true negative 0

Num. of true positive 1128 Accuracy 100%

Fig. 2 Sample of FCM + NN training for cheating tendency

Figure 3 shows the train-test split ratio of 80:20 for the dataset; therefore, 20 users will be used to assess the efficacy of FCM + NN in detecting cheating behavior. The 91.94% accuracy is based on a total of 211 observations, 194 of which are positive and 17 of which are negative.

Tables 1 and 2 shows the results of a FNN analysis of user cheating tendency for datasets with a train-test ratio of 80:20, 70:30, and 65:35. The highest accuracy training results, 100%, were found in the datasets with a train-test ratio of 80:20 and 70:30, while the lowest accuracy training results were found in the dataset with a train-test ratio of 65:35, at 99.98%. The maximum typical accuracy is 100%, represented by datasets 1, 2, and 3. However, the 80:20 train-test split yields the greatest accuracy (91.85%) in the sample. The 65:35 dataset, which is 85.21% accurate, and the 70:30 dataset, which is 80.82% accurate. Based on these findings, it appears that datasets with a train-test split of 80:20 have the highest average user cheating tendency.

Table 3 displays an overview of the typical simulation outcomes for NN, FCM, and FCM + NN. Based on these findings, FCM + NN appears to be the most effective method overall. Previous studies have shown that NN has limited training capabilities, which leads to poorer detection accuracy [9]. The FCM + NN, on the other hand, has the highest average accuracy across all datasets thanks to its ability to learn from clusters. Together, they help fix the problems that plague NN and FCM alone. Therefore, FCM + NN provides a high degree of accuracy across all three datasets.

testing_fcm_nn

Home Page   Training FCM+NN

Processing

Load Data	Load Net+Center
Testing	Reset

Accuracy

Num. of Data	211
Num. of true positive	194
Num. of true negative	17
Accuracy	91.9431%

	Gender	Year	CGPA	Grade	Class	Prepare	TC (Actual)	TC (Predicted)
1	0	3	3.66	1	0	3	1	1
2	0	3	3.66	0	0	3	1	1
3	0	1	0	1	1	2	1	1
4	0	3	3.63	1	1	1	1	1
5	0	3	3.63	1	1	1	1	1
6	0	3	3.66	1	1	2	1	1
7	0	3	2.85	0	0	1	1	1
8	1	3	2.88	1	1	1	1	0
9	1	2	0	0	1	3	0	0
10	1	3	2.9	1	0	2	1	0
11	1	3	3.53	1	0	5	1	0
12	1	3	2.89	1	0	2	1	0
13	1	3	3.17	1	0	2	1	0
14	0	3	2.79	1	0	1	1	1
15	0	3	2.79	1	0	1	1	1
16	0	3	3.1	1	1	3	1	1
17	0	3	3.92	1	1	3	1	1
18	0	3	3.26	1	0	2	1	1
19	0	3	3.26	1	0	2	1	1
20	0	1	0	1	1	3	1	1
21	0	1	3.52	1	0	1	1	1
22	0	3	3.66	1	1	2	1	1
23	1	3	3.24	1	0	3	1	0
24	1	3	3.24	1	0	3	1	0
25	1	3	3.19	1	0	2	1	0
26	1	1	3.34	1	1	3	1	1

Fig. 3 Sample of FCM + NN testing on user cheating tendency

Table 1 Demographic profile

		Frequency	Percentage (%)
Gender	Male	486	43.09
	Female	642	56.91
Age	Less than 20 years	613	54.34
	21–24 years old	489	43.35
	25–29 years old	16	1.42
	More than 30 years	10	0.89
Mode of study	Full time	1105	97.96
	Part time	23	2.04
Level of study	Diploma	682	60.46
	Degree	444	39.36
	Master	1	0.09
	PhD	1	0.09
Year of study	Year 1	231	20.48
	Year 2	575	50.98
	Year 3	275	24.38
	Year 4	47	4.17
Residence	Yes	887	78.63
	No	237	21.01
	Others		0.35

Table 2 FCM + NN analysis' summary for user cheating tendency

Ratio	80:20				70:30				65:35			
Train/Test	Training		Testing		Training		Testing		Training		Testing	
No.	e	%	e	%	e	%	e	%	e	%	e	%
1	10	100	10	78.2	10	100	10	100	10	100	10	97.63
2	20	100	20	100	20	100	20	86.26	20	100	20	78.67
3	50	100	50	100	50	100	50	64.46	50	100	50	97.63
4	100	100	100	100	100	100	100	54.5	100	100	100	100
5	183	100	183	91.94	183	100	183	50.54	183	99.76	183	54.5
6	247	100	247	100	247	100	247	97.63	247	100	247	100
7	451	100	451	100	451	100	451	98.56	451	100	451	79.62
8	684	100	684	78.2	684	100	684	99.53	684	100	684	61.14
9	772	100	772	91.94	772	100	772	75.36	772	100	772	91.47
10	978	100	978	78.2	978	100	978	75.36	978	100	978	91.47
Average	350	100	350	91.85	350	100	350	80.82	350	99.98	350	85.21
SD	351	0	351	9.95	351	0	351	1571.89	351	0.08	351	16.4
Min	10	100	10	78.2	10	100	10	54.5	10	99.76	10	54.5
Max	978	100	978	100	978	100	978	5054	978	100	978	100

Table 3 Summary of average simulation results

Training/Testing	Ratio	FCM + NN (%)
Training	80:20	100
	70:30	100
	65:35	99.98
	Mean	99.99
Testing	80:20	91.85
	70:30	99.98
	65:35	85.21
	Mean	92.35

5 Conclusion

Because it takes so much time to evaluate students' abilities, online examinations have become essential in modern times. Therefore, this issue can be addressed in the future studies of student behavior monitoring by employing a fuzzy c-means neural network approach. The fuzzy c-means approach is preferable since it accurately predicts students' propensity to cheat with a reasonable amount of enthusiasm.

References

1. N.I.B. Adnan, Z. Tasir, Online social learning model, in *2014 International Conference on Teaching and Learning in Computing and Engineering*, pp. 143–144 (2014)
2. B. Boitshwarelo, A.K. Reedy, T. Billany, Envisioning the use of online tests in assessing twenty-first century learning: a literature review. Res. Pract. Technol. Enhanc. Learn. **12**, 16 (2017)
3. M.R. Hameed, F.A. Abdullatif, Online examination system. Int Adv Res J Sci Eng Technol. **4**, 106–110 (2017)
4. D. Starovoytova, S. Namango, Factors affecting cheating-behavior at undergraduate-engineering. J. Educ. Pract. **7**, 66–82 (2016)
5. J. Ramberg, B. Modin, School effectiveness and student cheating: do students' grades and moral standards matter for this relationship? Soc. Psychol. Educ. **22**, 517–538 (2019)
6. L.A. Jensen, J.J. Arnett, S.S. Feldman, E. Cauffman, It's wrong, but everybody does it: academic dishonesty among high school and college students. Contemp. Educ. Psychol. **27**, 209–228 (2002)
7. L.C. Hensley, K.M. Kirkpatrick, J.M. Burgoon, Relation of gender, course enrollment, and grades to distinct forms of academic dishonesty. Teach. High. Educ. **18**, 895–907 (2013)
8. S.F.N. Sadikan, An initial framework of fuzzy neural network approach for online learner verification process. Int. J. Adv. Trends Comput. Sci. Eng. **8**, 185–189 (2019)
9. D. Pallavi, T.P. Anithaashri, Novel predictive analyzer for the intrusion detection in student interactive systems using convolutional neural network algorithm over artificial neural network algorithm, in *2022 4th International Conference on Advances in Computing, Communication Control and Networking (ICAC3N)*, pp. 638–641 (2022)
10. J.A. Trivedi, Voice identification system using neuro-fuzzy approach. Int. J. Adv. Res. Comput. Sci. Technol. **2**, 300–301 (2014)
11. F. Taher, R. Sammouda, Lung cancer detection by using artificial neural network and fuzzy clustering methods, in *2011 IEEE GCC Conference and Exhibition (GCC)*, pp. 295–298 (2011)
12. J. Jithish, S. Sankaran, A neuro-fuzzy approach for domestic water usage prediction, in *2017 IEEE Region 10 Symposium (TENSYMP)*, pp. 1–5 (2017)
13. M.H. Baeshen, M.J. Beynon, K.L. Daunt, Fuzzy clustering: an analysis of service quality in the mobile phone industry, in *Handbook of Research on Intelligent Techniques and Modeling Applications in Marketing Analytics* (IGI Global, 2017), pp. 40–61
14. C. Lakheyan, U. Kaur, A survey on web usage mining with fuzzy c-means clustering algorithm. Int. J. Comput. Sci. Mob. Comput. **2**, 160–163 (2013)

A Review on Maintenance Repair, Refurbish, and Overhaul for Reusable Launch Vehicle Management Process Toward Its Development

Hazariah Mohd Noh, Nur Syahira Sharif, Muhamed Roihan Yusoff, Haslinawati Besar Sa'aid, Puteri Nur Syaza Wardiah Raja Zainol, Haidah Hafni Ridzal, Izni Fauzan Mohamad Zainol, and Rita Zaharah Wan-Chik

Abstract This research reviews the application of reusable launch vehicles (RLV) technology in the maintenance, repair, refurbish, and overhaul (MRRO) operations. The current MRO of aircrafts, including structures, engines, and systems, will adhere to the original equipment manufacturer (OEM). This will follow in accordance with standard approval for aircraft and components maintenance organizations, Part 145 for aircraft maintenance organization (AMO). This research intends to study the reusable launch vehicle (RLV) maintenance program outline in adaptation to the current MRO industry for launch vehicles in Malaysia. The research method includes

H. M. Noh (✉) · N. S. Sharif · R. Z. Wan-Chik
Centre for Women Advancement and Leadership, Universiti Kuala Lumpur Malaysian Institute of Aviation Technology, Lot 2891, Jalan Jenderam Hulu, 43900 Dengkil, Selangor, Malaysia
e-mail: hazariah@unikl.edu.my

N. S. Sharif
e-mail: syahira.sharif@s.unikl.edu.my

R. Z. Wan-Chik
e-mail: ritazaharah@unikl.edu.my

M. R. Yusoff · H. B. Sa'aid · P. N. S. W. R. Zainol · H. H. Ridzal · I. F. M. Zainol
Aerospace Department, Universiti Kuala Lumpur Malaysian Institute of Aviation Technology, Lot 2891, Jalan Jenderam Hulu, 43900 Dengkil, Selangor, Malaysia
e-mail: mroihan@unikl.edu.my

H. B. Sa'aid
e-mail: haslinawati@unikl.edu.my

P. N. S. W. R. Zainol
e-mail: puterinursyaza@unikl.edu.my

H. H. Ridzal
e-mail: haidah.ridzal@e-serbadkgroup.com

I. F. M. Zainol
e-mail: izni.mohamad@e-serbadkgroup.com

© The Author(s), under exclusive license to Springer Nature Switzerland AG 2024
A. Ismail et al. (eds.), *Technological Frontiers and Sustainable Innovations*,
SpringerBriefs in Applied Sciences and Technology,
https://doi.org/10.1007/978-3-031-68751-8_5

an evaluation of the questionnaires given to the aviation and aerospace industries in Malaysia. This questionnaire was developed to provide industry insight into important considerations for the operations and maintenance of RLVs. The outcome of this study also addresses the parameters that can be used in the further analysis for the development on space-related activities, including the RLV and the feasibility study for the commercial launch platform in Malaysia.

Keywords Maintenance · Aviation · MRO · Aerospace · MRRO · RLV

1 Introduction

The current MRO of an aircraft (engine, avionics system, or airframe) will be following the original equipment manufacturer (OEM) servicing or repair station under the Part 145 (AMO) [1]. The maintenance and support system required by the aviation and aerospace industry companies is utilized primarily for corporate, commercial, and military aircrafts. The aviation industry definition of maintenance commonly includes all tasks required to restore or maintain an aircraft's systems, components, and structures in an airworthy condition. Three principal maintenance reasons include operational, value retention, and regulatory requirements [2]. Operational includes ensuring the aircraft is serviceable, reliable, and fit to fly. Value retention discuss in maintaining the current and future value of the aircraft by minimizing any physical deterioration during its life. In aviation MRO, the regulatory requirements are crucial where the condition and the maintenance of aircrafts are regulated by the aviation authorities such as the Civil Aviation Authority Malaysia (CAAM), i.e., the jurisdiction in which the aircraft is registered. The requirements establish standards for repair, periodic overhauls, and modification by calling for the owner or the operator to establish an airworthiness maintenance and inspection program to be carried out by certified individuals qualified to issue an airworthiness certificate (CoA). CoA will ensure a variety of the tasks such as (maintenance records, modifications, limited liability partnership (LLPs), configuration, release to service, repairs, recording of hours and cycles, mass, and balance). Details of the aircraft maintenance program are shown in Fig. 1.

The scheduled maintenance that includes the routine task card will be based on the development of the regular operation of an aircraft. This will require unscheduled, on-routine maintenance to troubleshoot for further investigating, repair the findings and discrepancies, or remove and restore faulty components. A requirement for unscheduled maintenance may result from scheduled maintenance tasks, pilot reports, or unforeseen events, such as high-load events, hard or overweight landings, tail strikes, ground damage, lightning strikes, or an engine over-temperature [3, 4].

Unlike aviation MRO, RLV MRRO first needs to be identified, and a detailed understanding of what RLV is. It is vital. RLV here will refer to a one-piece expendable rocket that might also achieve orbit with a single-stage and a completely reusable multistage vehicle that could be constructed. RLV refers to an utterly reusable

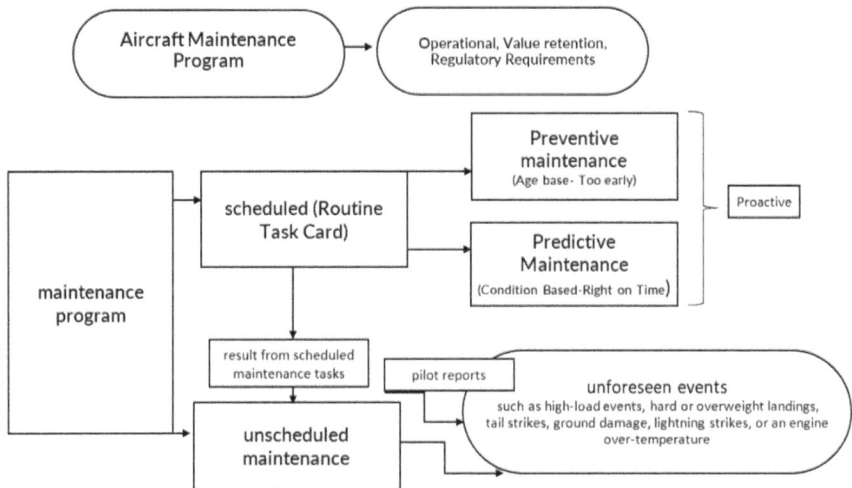

Fig. 1 Maintenance repair and overhaul (MRO) for aircraft maintenance program.

vehicle capable of achieving Earth orbit while carrying some useful payload and then returning [5]. A launch vehicle is a rocket-powered spacecraft that carries a craft past Earth's atmosphere. Today, practical launch vehicles have put human-crewed spaceflight shuttle missions, uncrewed space missions, and satellites into orbit since the 1950s [6]. Russia's Soyuz and Proton rocket launchers, as well as a range of modified military missiles, are among them; the Angara launcher family is also in the works. The Ariane V and Vega launchers are under European control [7]. The USA operated the space shuttle until it was retired in 2011. The Antares, Falcon, Atlas, and Delta disposable advocates are currently used in the USA [8].

In assessing the MRO and MRRO, the application of the launch vehicles, including the uncrewed spacecraft, single-use systems that are contionously monitored and used to improve throughout their prelaunch life cycle. For crewed spacecraft, the vital worry is crew safety. Space shuttles, for example, are thoroughly overhauled by NASA and the contractor personnel after every flight. Figure 2 shows the MRRO for the RLV [9].

The maintenance program for RLV focuses on system safety, reliability, maintainability, operability, and supportability [10]. A launch vehicle consists of additional rocket engines, power for those engines or devices brought in by fuel tanks, guidance, supervision, navigation, and command methods, freight, and a framework that holds all of these components together and to which additional engines or machines can be added for higher lift [11–13].

To bring in as much payload as feasible, launch vehicles are made as light as possible. A launch vehicle's components are mandatory to be tested as the vehicle is exposed to the stresses. It travels faster than the speed of sound and traverses the atmosphere, and the rocket engines operate under extreme pressure, compression, temperature, shock, tremor, and vibration [14]. RLV operations and maintenance

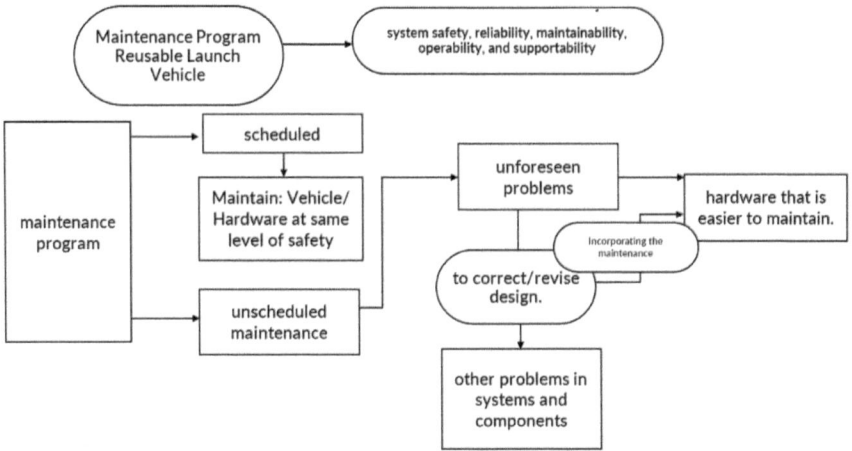

Fig. 2 Maintenance repair, refurbish, and overhaul (MRRO) for reusable launch vehicle

involve operations either with or without personnel aboard, including reusable rockets and orbiting RLVs [15, 16].

2 Methodology

The research method for this paper includes focusing on a comprehensive literature review to set up the foundation of the research. This includes the history and origin of launch vehicles and the people/country that pioneered the preliminary study. Understanding the functions of regulatory entities like the National Aeronautics and Space Administration (NASA), the European Space Agency (ESA), and the Malaysian Space Agency (MYSA) through its Space Exploration 2030 [17] is important in establishing the best framework for Malaysia. Interviewing key players in the MRO industry is also essential to gauge its presence in Malaysia in adaptation toward MRRO with references to the international aerospace industry.

MYSA and Space Technology Division are examples of the respondents discovering firsthand information regarding the actual status and process of aerospace MRO and the MRRO for RLV in Malaysia [18, 19]. Supporting the direction of the outcomes of this research can be found based on empirical evidence derived from the Spacevio publication for the feasibility study with the business climate in Malaysia with the results as "Space Ready" [20]. These questionnaires will also allow us to create the foundation of a feasibility study for an actual MRRO in Malaysia that caters to both aerospace and space.

3 Results and Discussion

The results based on the responses to the dedicated web-based questionnaires will be presented in this chapter. Working professionals in the Malaysian aerospace industry will be the criteria for the respondents. The data is tabulated and analyzed to determine the association between demographic variables such as the respondent's background, industry, work experience, area of expertise, and understanding of Space activities in Malaysia.

3.1 Industry Background

The varieties of the respondent's background include personal from aviation, aerospace and only less than 10% coming from other industries such oil and gas.

3.2 Area of Expertise

The questionnaires were given to the direct individual and were not open to the public. The expertise primarily comes from technical experts, management professionals, and safety personnel.

3.3 Awareness of Malaysia Space X-2030

From the results of question 6 in Sect. 2, most respondents are aware of the Malaysia Space Exploration 2030 Programs (Malaysia Space–X-2030. This leaves a positive note for both government and local companies in the aerospace industry since there must have been positive bilateral communication between the two parties. It is safe to assume that the Malaysian government's initiatives in creating a campaign for a more substantial aerospace presence are moving in the right direction.

3.4 Best Type of Launch in Malaysia

Looking at the question when the respondents were asked which type of launch setup is best for a country like Malaysia, most of the respondents chose ground launch; the reason behind this might be because of the strategic potential areas available land that Malaysia has, and the ground launch setup is the most and familiar by the respondent.

3.5 Malaysia's Capability of Having Its MRRO

67% of the respondents agreed that Malaysia has the capability to have MRRO.

3.6 Development of Reusable Launch Vehicle MRRO

The result for the "Development of Reusable Launch Vehicle MRRO" where the respondent's expert opinion was also being measured under the eight parameters, from which they were asked to individually assess whether it is essential to ensure the efficiency of a launch or not. The eight parameters are:

1. Space legislation
2. Financial capital
3. Manpower
4. Technical capabilities
5. Physical machinery
6. Time
7. Politics
8. Infrastructure.

Getting the perspective of industry players in the aerospace industry is very important in developing projects and initiatives. The correlation of the key factors will help the researcher create and understand the local content. It can create a platform to establish Malaysia as a stronghold of the aerospace/space industry in the Southeast Asia region. When respondents were asked which of the following affects the formation of RLV MRRO in a country with the same conditions as Malaysia, the majority chose financial capital as a critical element in creating an MRRO organization, followed by the geographical elements (infrastructure), technical capabilities of personnel and post holders, the manpower that is involved in doing the MRRO activities, respondents think that political environment of Malaysia is also a critical factor in the formation of MRRO, as well as proper space legislation, time, and physical machinery.

4   Conclusion

The concept of a RLV MRRO in Malaysia is new, but all earlier conclusions from previous discussions stated that it was economically nonviable. Any advancements in reduced-cost access to space promise to benefit the launch vehicle industry. Technology and capability development efforts across government and industry provide new tools for such recovery efforts.

The researcher's endeavor in this study aims to understand the Malaysia aerospace industry in developing RLV maintenance for commercial launch vehicle platforms.

MYSA's contributions to the aerospace industry and creating and leading the opportunities are crucial. Space Technology Division at Serba Dinamik also plays an essential role in creating the collaboration under New Space Economy Nexus (NSEN) that gathered the seven universities' partners in accelerating space activities through industries driven. The collaboration is yet to be well established, but exploring the currently available information with international support can enhance the further investigations of RLV MRRO readiness here in Malaysia.

Acknowledgements The authors thank the Gas Turbine Engine Research cluster of UniKL MIAT for providing all the necessary resources to contribute to this paper. Additionally, the authors appreciate the Space Technology Division for supporting this research work.

References

1. CAAM, *Maintenance Organisation Approvals*. CAAM Part 145 (2012). https://www.caam.gov.my/wp-content/uploads/2021/04/CAD-8601-Continuing-Airworthiness-of-Aircraft-CAAM-Part-M.pdf
2. Boeing 737 Aircraft Maintenance Manual, *Section 1 Introduction Maintenance, Scheduled Development, Program Checks, Maintenance Estimates, Maintenance Manhour Standards, Maintenance Performance Check* (Boeing, 2000)
3. IATA 2015, *Best Practices for Component Maintenance Cost Management* (2015). https://www.iata.org/contentassets/bf8ca67c8bcd4358b3d004b0d6d0916f/ac-leases-4th-edition
4. J. Moubray, Maintenance management: a new paradigm. CIM Bull. **94**(1055), 78–86 (2001)
5. M.A. Rampino, *Concepts of Operations for a Reusable Launch Vehicle* (Air Univ Maxwell AFB, Montgomery, AL, 1997)
6. E.M. Goodger, Jet fuels development and alternatives. J. Aerosp. Eng. **209**(2), 147–156 (1995). https://doi.org/10.1243/PIME_PROC_1995_209_281_02
7. J. Hospodka, Z. Houfek, Efficiency in carrying cargo to earth orbits: spaceports repositioning. MAD-Mag. Aviat. Dev. **4**(20), 6–9 (2016). https://doi.org/10.14311/mad.2016.20.01
8. V. Djurkovic, N. Milenkovic, S. Trajkovic, Dynamic analysis of rockets launcher. Tehnički vjesn. **28**(2), 530–539 (2021)
9. J.K. Tinoco, C. Yu, R. Firmo, C.A. Castro, M. Moallemi, R. Babb, Sharing airspace: simulation of commercial space horizontal launch impacts on airlines and finding solutions. J. Sp. Saf. Eng. **8**(1), 35–46 (2021)
10. J.T. Middendorf, J. Mendonca, *Reusable Launch Vehicle Operations and Maintenance Guideline Inputs and Technical Evaluation Report: Approval*, vol. 5 (Federal Aviation Administration (FAA), 2004)
11. S. Elbasuney, M. Gobara, M. Yehia, Ferrite nanoparticles: synthesis, characterization, and catalytic activity evaluation for solid rocket propulsion systems. J. Inorg. Organomet. Polym. Mater. **29**, 721–729 (2019)
12. A.M. Lipanov, V.E. Zarko, Survey of solid rocket propulsion in Russia, in *Chemical Rocket Propulsion*, ed. by L. De Luca, T. Shimada, V. Sinditskii, M. Calabro, Springer Aerospace Technology (Springer, Cham, 2017). https://doi.org/10.1007/978-3-319-27748-6_43
13. R. Tomanek, J. Hospodka, Reusable launch space systems. MAD-Mag. Aviat. Dev. **6**(2), 10–13 (2018)
14. N. Gayathri, A. Suhane, V.K. Khare, Maintenance strategy and its importance in rocket launching system—an Indian prospect, 6912–6918 (2016)
15. P. Eymar, F. Deneu, Reusable launch vehicles from a European point of view, in *Beyond the International Space Station: The Future of Human Spaceflight: Proceedings of the International Symposium*, Springer, Netherlands (2002)

16. J.L. Heldmann, A. Colaprete, D.H. Wooden, R.F. Ackermann, D.D. Acton, P.R. Backus, V. Bailey, J.G. Ball, W.C. Barott, S.K. Blair, M.W. Buie, LCROSS (Lunar crater observation and sensing satellite) observation campaign: strategies, implementation, and lessons learned. Space Sc. Rev. **167**, 93–140 (2012)
17. Malaysian Space Board Act 2020 (2020) CLJ Law
18. The Star, *Serba Dinamik to Set Up a New Subsidiary for Space Industry Venture in Q3* (2021). https://www.thestar.com.my/business/serba-dinamik-to-set-up-a-new-subsidiary-for-space-industry-venture-in-q3
19. The Sun Daily, *Gearing Up for Commercial Space Age: Malaysian Space Board Bill 2020* (2021). https://www.thesundaily.my/business/gearing-up-for-commercial-space-age-malaysian-space-board-bill-2020-CB8030730
20. Spacevio, *A Feasibility Study for Business Opportunities of Czech Based Space Companies in Malaysia-Space Ready*. Spacevio Publishing, 1–25 (2021)

Knowledge Management Issues in Higher Education Institutions: Still Unresolved Issues?

Suzana Basaruddin and Haryani Haron

Abstract In placing Malaysian education on the global map, the education system in Malaysia has undergone a series of changes in the early twenty-first century. The higher education institutions (HEI) were said facing difficulties in managing high expectations from their stakeholders, in ensuring the curriculums are always mapped to the industrial needs. The higher education authority has highlighted that curriculum review has not been given serious attention by HEI and this might give impact to the quality of graduates produced by HEI. The purpose of the qualitative study is to demonstrate the current discovery of knowledge management themes in HEI which emerge after close analysis from interviews using curriculum review as the context of the study. A qualitative approach using a single case study was employed in ensuring appropriate exploration of the context in selected case sites. Data collection method was conducted using interviews, before the analysis using the thematic approach took place. The themes then were analyzed and finally derived into four main issues in managing HEI knowledge that are (1) knowledge is scattered and isolated, (2) knowledge is inaccessible, (3) knowledge is unstructured, and finally (4) knowledge quality is unverified. It was found that this result matched with early studies of literature reviewed. The study contributed toward the understanding and conforming to the current knowledge management issues in HEI looking into curriculum review lens. This study might be the best approach for the HEI before moving into knowledge management approaches to overcome the knowledge discovered issues. This study could be expanded in the future by using multiple case studies and focusing on various contexts of knowledge in HEI.

S. Basaruddin (✉)
Universiti Kuala Lumpur, Malaysian Institute of Information Technology, Jalan Sultan Ismail, 1016 Kuala Lumpur, Malaysia
e-mail: suzana.basaruddin@unikl.edu.my

H. Haron
Faculty of Computer and Mathematical Sciences, Universiti Teknologi MARA, Shah Alam Campus, 40000 Shah Alam, Selangor, Malaysia
e-mail: harya265@uitm.edu.my

© The Author(s), under exclusive license to Springer Nature Switzerland AG 2024
A. Ismail et al. (eds.), *Technological Frontiers and Sustainable Innovations*,
SpringerBriefs in Applied Sciences and Technology,
https://doi.org/10.1007/978-3-031-68751-8_6

Keywords Knowledge management · Higher education institution · Curriculum review · Qualitative approach · Single case study

1 Introduction

The higher education institutions (HEI) in Malaysia are facing tremendous changes which started in the 1990s where researchers in the domain highlighted the importance of learning instead of teaching. Since then, teaching is not the only major activity in universities. HEI have included teaching and learning, research elements, focus or major, collaboration with industry, and services to communities and are hybrid or cross-disciplinary [1–5].

Authors of [6] and [7] highlighted that the main challenge in HEI is updating the content of courses offered in the HEI. The contents need to always meet the industrial level and needs [4, 8]. Agreed by and [9], all of this enforcement is to make sure graduates are competitive and have the skills required to enter the challenging industrial market.

The Malaysian Qualification Agency (MQA) has highlighted in its report [10] that the curriculum review has not been enforced accordingly. Improper knowledge management might prevent university members from getting the correct information and knowledge residing in the repository of the university. [11–15], [17] in their study of HEI knowledge management (KM) supports the KM scenario in HEI. Therefore, there is a need to study the current scenario of managing knowledge in HEI before finding an appropriate solution to ensure a better understanding of knowledge, thus allowing navigating of knowledge and promoting accurate knowledge retrieval is required.

The study examines the patterns in the meaning from the data inspected and collected and presented in the respondents' feedback and then analyzes the data into a group of themes and depicts the themes within relevant words (nouns and actions). This study aims at discovering the themes of HEI knowledge management that were derived after detailed analysis of data from interviews. These research contributions are not based on generalizations but rather on contextual findings and discoveries that are based on the qualitative research approach.

2 Methodology

Researchers initiated this study by understanding the issues of KM in HEI from the literature review. Then, researchers conducted a simple pilot study to confirm the issues. The investigation of the case site KM issues has been done through interviewing two respondents. They are programs' coordinators-cum-lecturer and have more than five years of working experience in HEI. The interview script was analyzed and researchers found out that the initial study was consistent with the

Table 1 Selected respondents

Respondents and unit	Relevant knowledge (respondent post)
RP1 **Information technology**	University operations knowledge (director)
RP2 **Record management**	University operations history knowledge (director)
RP3 **Library**	University corporate, academic and administrative knowledge (senior executive)
RP4 **Faculty**	Faculty academic and administrative knowledge (program coordinator)
RP5 **Faculty**	Faculty academic and administrative knowledge (program coordinator)
RP6 **Academic unit**	University academic and administrative knowledge (director)
RP7 **Academic unit**	University academic and administrative knowledge (deputy director)
RP8 **Academic unit**	University academic and administrative knowledge (deputy director)

literature review that revealed issues of KM in HEI. HEI was found to have been practicing the curriculum review process; however, there are some issues on ensuring that all academic members have the same understanding of the processes.

The selected case site is among established HEI in Malaysia. The academic unit manages the university curriculum review knowledge. As suggested by MQA, the unit has been implementing their SOP related to curriculum review and is actively promoting curriculum review processes via conducting continuous training for their academic staff.

Purposive sampling was selected in this study. This is because researchers need to select respondents of case sites that are keeping relevant knowledge and records of the curriculum review process. The respondents are selected from various top management posts from various unit (depicted in bold in Table 1) due to the relevant top-down view of processes. There are directors and deputy directors, senior executives, and program coordinators.

Five units of the university were selected before further decisions on the respondents. The criteria of the respondents are that they have been serving the university for at least five years and have been involved in operations and management of the unit/department. Finally, eight respondents were named as RP1 until RP8. Details of the respondents are mentioned in Table 1.

2.1 Interview Procedure

The interview process has to be planned and managed before the interview session takes place. Researchers use an interview protocol in ensuring the interview sessions are all set for a smooth process and able to achieve the interview aims.

Table 2 Responses of respondents after analysis

Responses of respondents	
"There are various systems in various platforms in these academic institutions and have been developed by different units or departments and being used and maintained by another different units or departments. There is no master list showing all the systems and information or knowledge that is being managed. This includes knowledge related to the curriculum." (RP1)	Decentralize knowledge Isolated knowledge Scattered knowledge Inconsistency knowledge process Minimum review Unstructured knowledge, Less monitoring process Less monitoring process
"Every unit or department such as the curriculum unit and faculty are managing their own information and knowledge." (RP2, RP3)	
"We share all important resources at our websites and electronic systems." (RP4, RP5, RP6, RP7, RP8)	
"Any procedures, rules or regulations have been distributed through the representative or committees and related unit or department. But some information does not reach the faculty members leading to inconsistency of curriculum implementation or execution." (RP4, RP5, RP6, RP7, RP8)	
"Curriculum review shall be done at maturity of the student cohort or upon instruction to standardize any part of the documents." (RP4, RP5, RP6, RP7, RP8)	
"Curriculum design, development and storage was done at faculty level. But some of the documentation is not the latest and some is untraceable. Most of the knowledge resides as a tacit knowledge among previous faculty members that were in any positions before." (RP7, RP8)	
"The coordinator and administration office should have all the latest copy of all curriculum related. Each lecturer must get the latest curriculum from either one of these people, but there is no counter check on this. This means the lecturer might have used the obsolete or not the latest curriculum." (RP5, RP6, RP8)	

2.1.1 Interview Protocol

The preparation of interview consisted of four main sections that are:

1. Pre-interview

It is prepared before the interview. The pre-interview scripts contain the information, steps, and process before the interview session takes place, for example, the session arrangement and setting of the interview (duration of the interview session).

2. Interview session

It is used just before the interview takes place. Information collected is information about the date, time, venue, respondent name, post, and department or unit.

3. Interview scripts

It contains the questions of the interview. This has been the guideline for the interview session, in order to make sure the respondent gives appropriate answers relevant to the interview session. Interviewer uses probing to achieve the interview aim. All the answers or responses have been jotted down in the forms prepared.

4. Post-interview Scripts.

Post-interview scripts contain scripts to be used at the end of the interview session. Interviewer concludes the session and verifies the notes taken during the interview.

The interview sessions conducted were based on the set of questionnaires prepared in the interview protocol. Interviewer may use various approaches as long as all the listed questions have been covered. Probing was highly encouraged to get into appropriate answers. The lists of questions prepared are based on four HEI curriculum review knowledge that are:

1. What are the processes involved in this department/unit?
2. How are the documentation processes?
3. What are the related issues or problems?
4. What is the relevant knowledge in HEI that supports the HEI curriculum review process?

2.1.2 Interview Analysis

When the interview session ends, scripts from the interviews have been revisited. Points jotted down have been gone through thoroughly. It was in the mixed language Bahasa Melayu and English. The Bahasa Melayu was paraphrased in English. Responses (facial expressions) noted down also were translated into appropriate expressions. Example of the process is depicted in Fig. 1.

The analysis method used in the study is the content analysis. In this process, each paragraph of the script has been assigned with a meaning of full theme or themes according to the meaning of sentences in the paragraph. At least two rounds of this

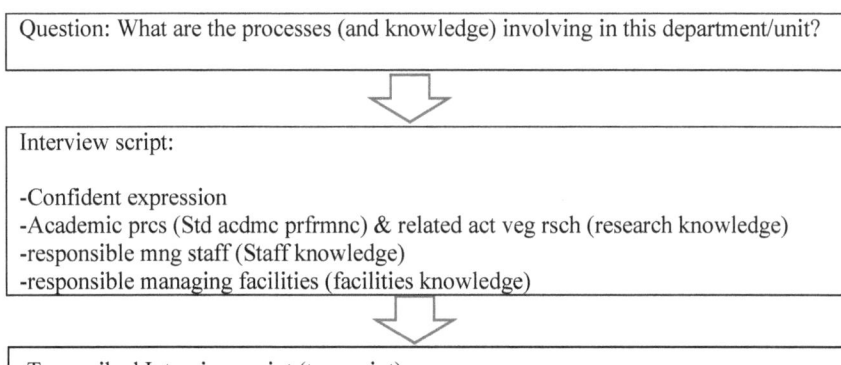

Fig. 1 Example of transcribing

process were carried out. One round of the process has been done by researchers; another round has been revised by experienced researchers' supervisors.

All those themes then were reviewed to group similar themes by paraphrasing, generalizing, removing, or abstracting the themes. This process was repeated for three times and has been discussed by the expert of curriculum review that is the academic unit. Finally, the final themes remain as the outcome of the analysis. Some significant responses of respondents before finalizing the themes are depicted in Table 2.

3   Result and Discussions

The aim of this study is to confirm issues of KM in HEI. This is because it is critical to solve this issue before any other KM activity should take place in the HEI. Final outcome of the analysis revealed four main themes that are:

1. *Curriculum review knowledge is scattered and isolated.*
2. *Some curriculum review knowledge is inaccessible.*
3. *Curriculum review knowledge is unstructured.*
4. *Most curriculum review knowledge quality is unverified.*

The analysis also found that the most curriculum review knowledge being used, reused, and shared is from the academic unit. Thus, this study recommended the academic unit to do proper knowledge audits related to curriculum review and take necessary action to integrate and centralize, provide more accessibility options, revise the structuring, and include verification process in the current knowledge flow.

4   Conclusions

Academic institutions have been put responsible for ensuring that the curriculum taught in the HEI is at the best to produce knowledgeable and competitive graduates in the academic world and industry. These high expectations must be fulfilled by HEI by ensuring the latest and updated curriculum is embedded and applicable by the graduates. Therefore, the curriculum review process must be made at the fingertips of academicians, to always be aware of when and how to update the curriculum when necessary.

Findings in this study have confirmed the issues of HEI knowledge management are still unresolved issues, looking into the lens of curriculum review. Some suggestions for the main stakeholder have been delivered so that the curriculum review knowledge will be improved. Without these initiatives, HEI may not be able to keep up with stakeholders' expectations especially when the role of academicians has been increased to many more challenges in achieving excellence. The study also provides an indicator to HEI that in keeping up with the latest industrial revolution and globalization issues, knowledge management has become more pertinent and significant.

References

1. M.G. Mohayidin, M.N. Kamaruddin, M.I. Margono, The application of knowledge management in enhancing the performance of Malaysian universities. EJKM **5**(3), 301–312 (2007)
2. T.A. El-Badawy, Y. Hashem, The impact of social media on the academic development of school students. Int. J. Bus. Adm. **6**(1), 46 (2015). https://doi.org/10.5430/ijba.v6n1p46
3. N. Hussein, S. Omar, F. Noordin, N.A. Ishak, Learning organization culture, organizational performance and organizational innovativeness in a public institution of higher education in malaysia: a preliminary study. Proced. Econ. Financ. **37**, 512–519 (2016). https://doi.org/10.1016/S2212-5671(16)30159-9
4. Chandran, D.P.K., Kooi, Y., Harizan, M.H.M., Kooi, C.C., Hoy, T.T. Foong, C.K.: Success story of collaboration between Intel and Malaysian universities to establish and enhance teaching and research in electronic packaging. in 34th International Electronic Manufacturing Technology (IEMT) Symposium, pp. 1–6, (2010). https://doi.org/10.1109/IEMT.2010.5746684
5. L. Arokiasamy, M. Ismail, A. Ahmad, J. Othman, Background of Malaysian private institutions of higher learning and challenges faced by academics. J. Int. Soc. Res. **2**(8), 60–67 (2009)
6. Daud, N.: Quality of work life and organizational commitment amongst academic staff: empirical evidence from Malaysia. In 2010 International Conference on Education and Management Technology (ICEMT 2010), pp. 271–275, (2010). https://doi.org/10.1109/ICEMT.2010.5657657
7. B.L. Hall, R. Tandon, Decolonization of knowledge, epistemicide, participatory research and higher education. Res. All **7**(1), 6–19 (2017). https://doi.org/10.18546/RFA.01.1.02
8. Mohammad, M.F., Abdullah, R., Jabar, M.A., Nor, R.N.H.: Establishing knowledge management model of quality management systems for higher education institutions. In Global Perspectives on Quality Assurance and Accreditation in higher education Institutions, IGI Global pp. 90–118, (2022). https://doi.org/10.4018/978-1-7998-8085-1.ch006
9. MQA (2017) Malaysian Qualifications Framework (MQF) Version 2.0 Draft Stakeholders' Consultation. http://www.mqa.gov.my/PortalMQAv3/dokumen/maklum%20balas/MQF%20V2%20DRAFT2.pdf. Accessed 25 Dec 2017
10. Suhaimee, S., Bakar, A. Z. A.: Knowledge management implementation in Malaysian public institution of higher education, In Proceedings of the Postgraduate Annual Research Seminar, pp. 47, (2005). https://www.academia.edu/9337223/Knowledge_Management_Implementation_In_Malaysian_Public_Institution_of_Higher_Education Accessed 20 Oct 2020
11. M.B. Yaakub, K. Othman, A.F. Yousif, Knowledge management practices in Malaysian higher learning. Int. J. Educ. Res. **2**(1), 1–10 (2014)
12. S. Basaruddin, H. Haron, S.A. Noordin, Managing knowledge in academic institutions using corporate taxonomy. Int. J. Circuit Theory Appl. **10**(07), 87–95 (2017)
13. Chipeta, G.T. Chawinga, W.D.: Knowledge management capability in higher education: The Case of Lecturers at Mzuzu University, Malawi. Managing Knowledge and Scholarly Assets in Academic Libraries, IGI Global, pp. 302–333, (2017). https://doi.org/10.4018/978-1-5225-1741-2.ch015
14. O.F. Al-Kurdi, R. El-Haddadeh, T. Eldabi, The role of organisational climate in managing knowledge sharing among academics in higher education. Int. J. Inf. Manage **50**, 217–227 (2020). https://doi.org/10.1016/j.ijinfomgt.2019.05.018
15. M.Y. Cheng, J.S.Y. Ho, P.M. Lau, Knowledge sharing in academic institutions: a study of multimedia university Malaysia. EJKM **7**(3), 313–324 (2009)

Malaysia Space Race: A Study on Commercial and Military Spaceports

Hazariah Mohd Noh, Aishy Rania Sofea Azrul, Muhamed Roihan Yusoff, Haslinawati Besar Sa'aid, Puteri Nur Syaza Wardiah Raja Zainol, Rita Zaharah Wan-Chik, and Mohd Norsyamim Samandi Marguna

Abstract Spaceports are often developed in geopolitically stable regions that benefit the country and the user. Numerous spaceports are situated in the most physically optimal regions open to operators. These regions are distinguished by geographical features such as proximity to the equator, possibilities for launching in the direction of the east or near the east, and favorable environmental factors. There are now several commercial and military spaceports in various stages of development throughout the globe. This paper proposes a research plan to study the currently available commercial and military spaceports worldwide and simultaneously evaluate a spaceport for its essential characteristics. The researcher will propose the research plan using the systematic literature review (SLR), where all the documents regarding this study will

H. M. Noh (✉)
Centre for Women Advancement and Leadership, Universiti Kuala Lumpur Malaysian Institute of Aviation Technology, Lot 2891, Jalan Jenderam Hulu, Jenderam Hulu, 43900 Dengkil, Selangor, Malaysia
e-mail: hazariah@unikl.edu.my

A. R. S. Azrul · H. B. Sa'aid · P. N. S. W. R. Zainol · R. Z. Wan-Chik · M. N. S. Marguna
Aerospace Department, Universiti Kuala Lumpur Malaysian Institute of Aviation Technology, Lot 2891, Jalan Jenderam Hulu, Jenderam Hulu, 43900 Dengkil, Selangor, Malaysia
e-mail: aishy.azrul21@s.unikl.edu.my

H. B. Sa'aid
e-mail: haslinawati@unikl.edu.my

P. N. S. W. R. Zainol
e-mail: puterinursyaza@unikl.edu.my

R. Z. Wan-Chik
e-mail: ritazaharah@unikl.edu.my

M. N. S. Marguna
e-mail: syamim.samandi@e-serbadkgroup.com

M. R. Yusoff
Center of Helicopter and Aircraft Maintenance Professionals, Universiti Kuala Lumpur Malaysian Institute of Aviation Technology, Lot 2891, Jalan Jenderam Hulu, Jenderam Hulu, 43900 Dengkil, Selangor, Malaysia
e-mail: mroihan@unikl.edu.my

© The Author(s), under exclusive license to Springer Nature Switzerland AG 2024
A. Ismail et al. (eds.), *Technological Frontiers and Sustainable Innovations*,
SpringerBriefs in Applied Sciences and Technology,
https://doi.org/10.1007/978-3-031-68751-8_7

57

be studied and evaluated for detailed information and requirements for a spaceport. The research outcome will be the research design and methods, so the focal point of the outlook on geopolitical factors, history, and the components related to spaceports can be further investigated.

Keywords Spaceport · Commercial · Military · Malaysia

1 Introduction

A spaceport, commonly referred to as a cosmodrome, is a location where spacecraft are launched. In terms of its capacity to house ships and airplanes, it is analogous to a seaport or an airport. For a very long time, the terms "spaceport" and "cosmodrome" have been used to refer to establishments that can send spacecraft into Earth orbit or on interplanetary journeys [1–3]. The expressions "cosmodrome" and "spaceport" are frequently used synonymously. However, the expression "spaceport" is also frequently used to describe the sites of launch pads for rockets that are exclusively utilized for suborbital flights. This is due to the fact that the term "spaceport" is frequently used to refer to brand-new and envisioned launch pads for suborbital human flights [4–6].

This is the case mainly if space stations are intended to serve as spaceports. A rocket launch station is any location from which rockets are launched. There is a chance that one or more pads will be released, along with locations suitable for a portable pad launcher [7]. It is frequently encircled by a sizable safety zone and is also known as a rocket or missile range. The range represents both the expected path taken by launched rockets as well as the potential landing zone for specific rocket parts. Tracking stations are occasionally set up in the range to monitor the path of the launches [8, 9].

Major spaceports typically have many launch complexes or rocket launch locations that may be altered to accommodate various launch vehicle types. Major spaceports often have several launch complexes. The distance between these locations can be kept relatively large for safety purposes. For liquid-propellant launch vehicles, appropriate storage and, under some circumstances, production facilities are needed. On-site solid propellant processing is also prevalent [10–12].

More than only supporting the infrastructure for rocket launches is provided by spaceports. Customers and the aerospace industry can benefit from a number of services a spaceport offers. These include:

(a) Aerospace design and manufacturing capabilities to support launch vehicles and payload suppliers.
(b) Data transfer during a launch or test and telemetry and range support services for secure launches.
(c) Additional services include propellant supply and storage, secure facilities, weather monitoring, lightning protection, and more. Facilities for payload

processing and integration to assist satellite development, production, and testing.

The R&D of the Department of Defense, academic institutions, and private businesses can benefit from the same production, integration, and engineering services. A spaceport uses a variety of transportation methods where large rocket parts like solid boosters and fuel tanks are frequently manufactured off-site. Air, land, ship, and train are frequently distributed to these parts. Commercial spaceports are aided in their launch by local transportation infrastructure. Spaceports provide museums, tours, education, retail stores, and training for astronauts and pilots [13].

2 Research Methodology and Proposed Spaceport Investigation

Various commercial and military spaceports are now being built in different parts of the world. This study tries to determine the precise number of spaceports for commercial and military uses that have been built around the world. The researcher will consider geography, history, and operational and dormant spaceports when conducting this study. In this analysis, commercial spaceports converted to government spaceports were also included. Figure 1 illustrates this.

Malaysia, meanwhile, has not yet begun construction of a spaceport. The Malaysian aerospace sector may benefit from better planning in the future as our country is currently only in the study phase for the spaceport. Here, the priority spaceport factors will be assessed in order to establish whether Malaysia qualifies for a spaceport based on the study's consideration of those aspects.

This proposed investigation goes into great detail on each process, each phase, and each method and emphasizes the importance of choosing the best approach. The chapter concludes by suggesting that the techniques used to analyze the data must also be described. The following are a few of the subjects covered in this research plan:

(I)   Research design and research flow
(II)   Research method
(III) Method of collecting the data
(IV)   Systematic literature review
(V)   Expected outcomes

2.1 Research Design

The following diagram outlines the steps needed in doing this research. Figure 2 illustrates the data collection procedure that was used to meet the investigation's goals.

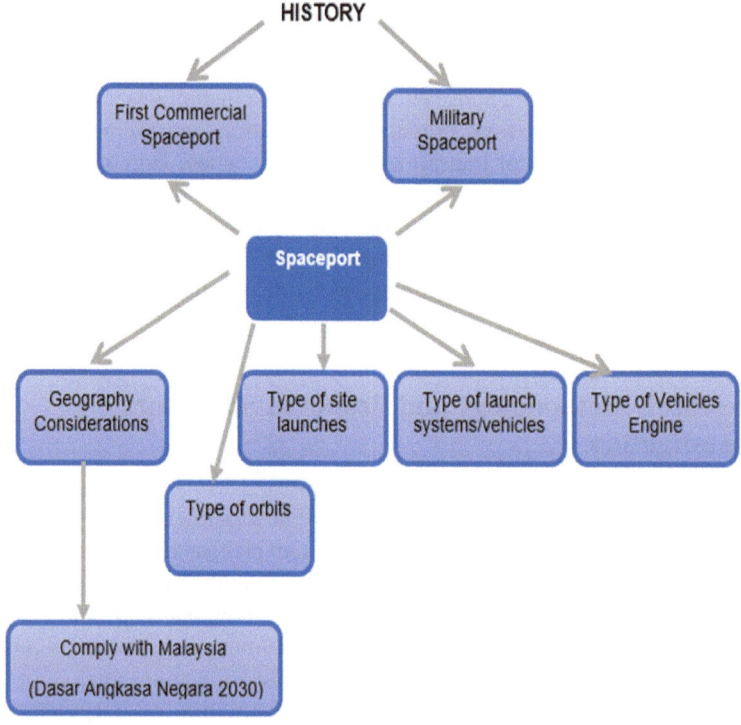

Fig. 1 Spaceport proposed study for commercial and military

In order to lay the groundwork for this research, preliminary studies must conduct a complete literature analysis that covers topics such as the history of spaceports, the origin of vehicle launches, and the global distribution of spaceports. Establishing the appropriate framework for Malaysia to measure its presence and determining if our nation is eligible to have a spaceport base based on those elements requires understanding the priority aspects of spaceports. It is important to understand the current availibility of spaceport for both commercial and military in ensuring its research integration and interpretation review as in (Fig. 3).

Finding, analyzing, and summarizing prior academic research on a specific topic or issue is the goal of a thorough, well-structured research endeavor known as a systematic literature review. A preset search plan is used to conduct systematic reviews, and the criteria are fully stated before the review is carried out. Other scholars are allowed to duplicate and repeat this thorough and open search [14, 15].

A search strategy focused on or targeted at a specific topic or question must be carefully designed for systematic reviews. The review highlights the types of information sought out, assessed, and reported on within predetermined time frames. Everything, including the search terms, search strategies (including database names, platforms, and search dates), and limits, must be considered throughout the evaluation.

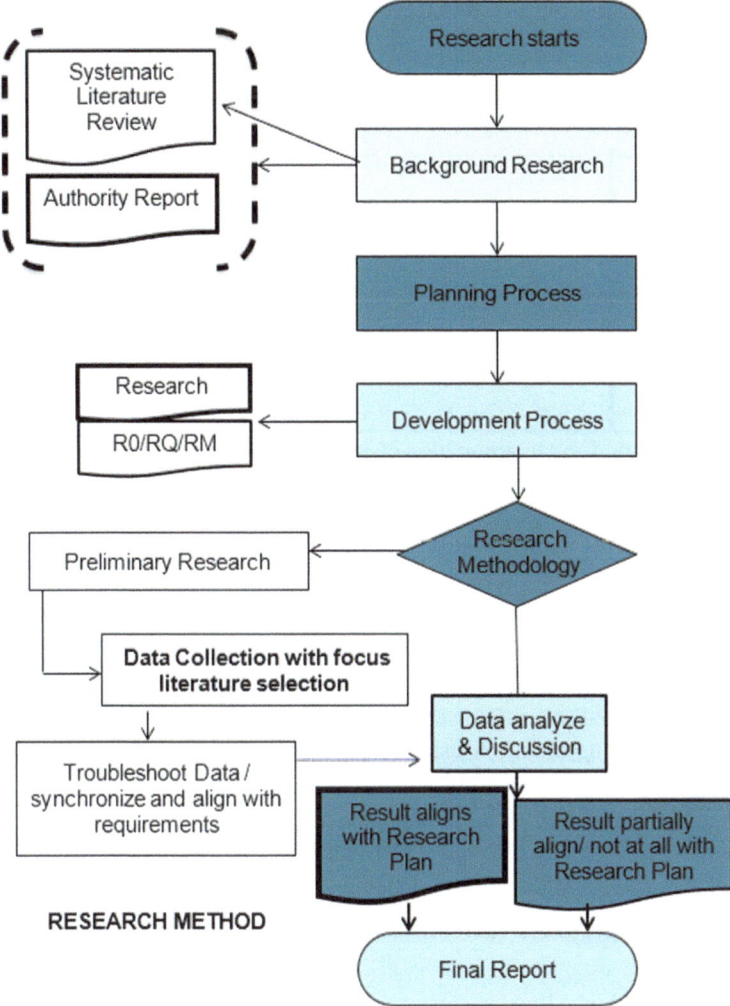

Fig. 2 Proposed research method

3 Malaysia Space Industry

The nation can meet its current need for space capabilities with wise international collaborations. However, Malaysia will eventually require a strategy to become a leader in space's strategic and practical use. The Dasar Angkasa 2030 policy serves as the initial step in strategically and effectively coordinating the country's space operations. Malaysia will not be forced to take part in expensive, risky space projects or unreasonable space exploration under the National Space Policy. Existing national policies will continue to serve their intended purpose and play their current role.

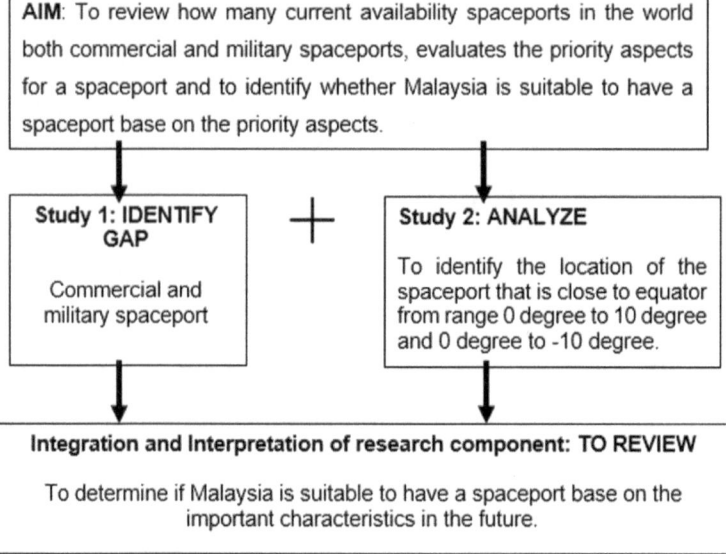

Fig. 3 Research integration and interpretation for spaceport

Instead, the policy recognizes the need for access to talents to enhance public services and national security to safeguard national interests. It will also serve as a roadmap for deciding Malaysia's space development to properly utilize Malaysia's space capabilities, strategically coordinate priorities and commitments in space, and provide enough resources [16].

3.1 Technological Advancement and National Space Policy

Thanks to technological advancement, strategic applications of space technology, particularly those pertaining to the defense industry, are progressing. Complementary technologies like mobile services enable access to previously unavailable technology and applications that can be used for social, environmental, and economic objectives. Maps are now instantly available, unlike in the past when they could only be found on paper. Both beneficial and detrimental outcomes could result from this. Malaysia must consequently manage the current wave of technological advancement and application development.

The nation's social, environmental, and economic health depends on our capacity to use space. It has become increasingly vital to rely on this space system, particularly for fundamental yet essential tasks like safety. Because of this, mastering space has to be done in a group. Space capabilities provide a supplement to and support current national policies, helping them achieve their goals.

The National Space Policy follows the coordination of domestic and international space issues. Malaysia must ratify international space-related treaties and agreements to demonstrate that it is a responsible government while conducting space-related operations. In order to conduct space affairs at the national level, particularly those relating to the peaceful exploration and use of space for the benefit and advantage of humans, governments are required to put into practice the principles outlined in the United Nations (UN) Outer Space Treaty [17].

4 Conclusion

This paper is expected to give a general overview of the proposed plan for a spaceport study, and by the outcomes of the studies, it would be possible to say whether Malaysia can make the most of its location to build a spaceport. This is because of this analysis of spaceport locations in close proximity to the equator pinpoints. This stance will unquestionably be advantageous for those nations with the ability to use space. Further spaceport review studies for the Malaysian context could analyze these potential benefits and quantify them in terms of economic, scientific, national security, environmental, and technological impact. It could also examine the potential costs associated with the development and operation of a spaceport, including infrastructure costs, environmental impacts, and regulatory compliance. Ultimately, such a study could help inform decisions about whether and where to establish a spaceport, as well as how to optimize its design and operation to maximize the benefits and minimize the costs.

Acknowledgements The author wishes to express his gratitude to UniKL MIAT's Gas Turbine Engine Research cluster for giving all the resources required for this paper. The Space Technology Division is also to be thanked for helping to support this investigation.

References

1. E. Seedhouse, Spaceports: a definition and brief history, in *Spaceports Around the World, A global Growth Industry*. Springer Briefs in Space Development. (Springer, Cham, 2017) https://doi.org/10.1007/978-3-319-46846-4_2
2. Space Foundation The space briefing book (A Reference Guide to Modern Space Activities). (2019) https://www.spacefoundation.org/briefingbook/
3. M. Jenkins, Strategic geographical points in outer (2021) space. https://www.thespacereview.com/article/4273/1
4. D. Wright, L. Grego, L. Gronlund, *The physics of space security (A reference manual)* (American Academy of Arts and Sciences, Cambridge, 2005)
5. What are the national seismic hazard maps for Japan? (2012) Japan Seismic Hazard Information Station. https://www.j-shis.bosai.go.jp/en/shm
6. M. Dachyar, H. Purnomo, Spaceport site selection with analytical hierarchy process decision making. Indian J. Sci. Technol. **11**(10), 1–8 (2018)

7. B. Gulliver, G. Finger, Can your airport become a spaceport? The benefits of a spaceport development plan, in: *48th AIAA Aerospace Sciences Meeting Including the New Horizons Forum and Aerospace Exposition* (2010)

8. R.L. Christenson, D.R. Komar, Reusable rocket engine operability modeling and analysis, National Aeronautics and Space Administration (1998)

9. G. Letchworth, X-33 reusable launch vehicle demonstrator, spaceport and range. in: *AIAA Space Conference and Exposition* (2011)

10. E. Seedhouse, Spaceport America, in: *Virgin Galactic. Springer Praxis Books*. (Springer, Cham, 2015) https://doi.org/10.1007/978-3-319-09262-1_5

11. S. Landeene, L. Gomez, A. Prescott, B. Ziarnick, Spaceport America: the world's first purpose-built commercial space. in: *AIAA SPACE Conference and Exposition* (2008)

12. G. Finger, J. Kercsmar, R. Hammett, D. Keller, P., Endicott, B. Gulliver, State spaceports-If you build it, will they come? in: *AIAA Space Conference & Exposition*

13. R. Babb, (2019) Building a 21st century multi-user commercial spaceport: Development and application of the spaceport readiness level scale. Doctoral Dissertations and Master's Theses. https://commons.erau.edu/edt/478

14. Roberts, G. Thomas (2019) Spaceports of the world. Center for Strategic & International Studies, (2007)

15. J. Memon, M. Sami, R.A. Khan, M. Uddin, Handwritten optical character recognition (OCR): A comprehensive systematic literature review (SLR). IEEE Access: 142642–142668. (2020) https://doi.org/10.1109/ACCESS.2020.3012542

16. R. van Dinter, B. Tekinerdogan, C. Catal, Automation of systematic literature reviews: a systematic literature review. Inf. Softw. Technol. (2021). https://doi.org/10.1016/j.infsof.2021.106589

17. A. Farhani, I.S. Chandranegara 2030 Ministry of energy, science, technology, environment and climate change. Agensi Angkasa Negara (ANGKASA) (2017)

An Insight into Logistics Management and Practices for Non-logistician

Zawiah Abdul Majid, Nor Aida Abdul Rahman, and Nurhayati Mohd Nur

Abstract An insight into logistics management and practices persists to be essential among researchers, educators, logistics service providers, and everyone engaged in supply chain management. However, the multi-definition and provision of logistics management practices are perceived inversely due to variations in school of thought that habitually is misrepresented. Therefore, this paper enriches the knowledge of logistics management practices for logisticians as well as non-logisticians. Logistics sectors are often presumed to be monopolized by males. However, women's roles as mothers, for example, in managing routine household without realizing they are an unseen logistician too. The rational understanding of this multi-definition will improve communication among logisticians as well as for non-logisticians. In addition, efficient logistics management practices are crucial to improve productivity leading to profitability. Women involvement is vital to improve their reputation, in addition to tackling the gender imbalance. Fostering and empowerment of women in logistics management and practices will be beneficial in improving their competency. Tag along with innovation and technology investment for logistics management and practices toward business sustainability.

Keywords Women · Logistics · Supply chain management (SCM) · Technology · Sustainability

Z. A. Majid (✉) · N. M. Nur
Universiti Kuala Lumpur, Malaysian Institute of Aviation Technology, Lot 2891 Jalan Jenderam Hulu, 43800 Dengkil, Selangor, Malaysia
e-mail: zawiah@unikl.edu.my

N. M. Nur
e-mail: nurhayatimn@unikl.edu.my

N. A. A. Rahman
Universiti Kuala Lumpur, Malaysian Institute of Aviation Technology, Persiaran A, Off Jalan Lapangan Terbang Subang, 47200 Subang, Selangor, Malaysia
e-mail: noraida@unikl.edu.my

© The Author(s), under exclusive license to Springer Nature Switzerland AG 2024　　65
A. Ismail et al. (eds.), *Technological Frontiers and Sustainable Innovations*,
SpringerBriefs in Applied Sciences and Technology,
https://doi.org/10.1007/978-3-031-68751-8_8

1 Overview

In this chapter, the multi-definition and provision on logistics management and practices will be illustrated. Henceforward, this clarification could enhance logistician and non-logistician understanding about the multi-definition of the logistics concept. Thus, this complication could avoid missunderstanding, preserving decent relationships in the direction of business sustainability. Logistics management and practices continue to be crucial among researchers, educators, service providers, and stakeholders engaged in supply chain management [1]. Consequently, the need to define logistics management and practices thoroughly is prerequisite to avoid misscommunication in damaging the business reputation. Logistics presents a substantial role in boosting a country's economy and industry growth. Logistics management is part of the process in the supply chain that concentrates on planning, implementing, and controlling efficiently, effectively the goods movement and storage, services, and information from the origin to consumer [2]. The term logistics can be defined as "the process of strategically managing the movement and storage of materials, parts, and finished inventory from suppliers, through the firm and on to customers" [3]. Transport is a crucial component of integrated logistics management. An established transportation system will warrant an improvement in reducing the cost of operation and in addition to a higher service quality. In other words, transport is a component of logistics that highlights the undertaking of commodities in the logistics along the supply chain [4].

2 Logistics Management and Practices

In the beginning of the past decades, this movement of goods is recognized as physical distribution. Significant changes as the requirement in logistics area turns out to be crucial and evolved as logistics management. The evolution and innovation of logistics service providers (LSP) can be seen through the growth of 1PL to 10PL [5]. In other words, the customer's demand is increasing and LSP needs to add value for their customers satisfaction to endure sustainability. Companies identified and progressed to build effective logistics management from the supply chain management perspective.

2.1 Logistics Service Providers

Logistics service providers are generally well-known as third party (3PL) and the people involved in LSP are categorized as logisticians. Predominant roles of logistician for planning, managing, and controlling logistics management and practices specifically transportation, warehousing, storage, inventory management, packaging, labeling, labor supply, and as requested by their customers [6]. LSP acts on behalf of their customers (manufacturers/exporters, factories/traders/consumers/importers or more) integrating operations such as coordinating import/export booking, land transportation including by rail or road, by air or by water including by river or sea, and through pipeline). Including shipping arrangements, customs documentation, cross-docking, port clearance, other government agencies, permit/fumigation, and other required services. LSP must be innovative to foreseen issues and challenges of globalization, IR 4.0 era, and the rapid changes of consumer preferences. In logistics rising rivalry, the selection of truthful LSP is prerequisite to coordinate on behalf of their customer toward the seven rights (7Rs); right customer, product, time, place, condition, quantity, and cost [2].

3 Supply Chain Management

Supply chain management is regarding movement of goods from the origin to end-user and logistics management is part of the SCM [3]. It is vital for companies to outsource and engage LSP to focus on their core-proficiency business. Interested parties that appoint LSP are companies involved in farm/exporters, factories/manufacturers, traders, consumers/importers, and others stakeholder in the supply chain management.

4 Modes of Transportation

Transportation services are from origin to destination using selected mode of transport such as by land transportation including by rail or road, by air or by water including by river or sea, pipeline, and inter-modal transportation. The range and methods of transport are based on the movement of goods pertaining to customers' demand. Priority is toward delivery on time with cost-effective, safe, right quality, and quantity of commodity purchased.

5 Multi-Definition of Logistics Practices

The multi-definition of logistics that are commonly referred to as below:

No	Type of logistics	Definition
1	Military logistics	The history of logistics begins through the military's demand on provisions of arms, ammunition, and ratio movement toward military centers. Dealing with staff evacuation and hospitalization support for health and medical services, distribution, maintenance evacuation, design and growth, motion, procurement storage, building maintenance operation, and material disposal
2	Humanitarian logistics	The definition of is to plan, implement, and control efficiently and effectively the movement of goods and storing including warehousing from the origin to end consumer for purpose of relieving the suffering and vulnerable people. Emphasis on aiding people in need, easing the suffering and pain of those impacted by tragedies with offering budget-friendly to the essential destination
3	Halal logistics	In upholding Halal, truthfulness namely Halal integrity is the main objective of Halal logistics in the movement of Halal products from origin to the end-user [7]. The Halal product, whether finished good, semi-finish, or raw material must be secured and remain Halal all the way in the Halal supply chain. The Halal logistics is definition "a process of planning, implementing, and managing the efficient, seamless flow and storage of Halal certified products (raw materials, semi-finished, or finished good) from the origin to the final consumption ensuring full Syariah compliance [8]." Certified Malaysia Halal Logistics Standard: MS2400-1-2010 Distribution, MS2400-2-2010 Warehousing and MS2400-3-2010 Retailing [9]
4	Cold chain logistics	Is a network of temperature control commodities, cold stores, refrigerators, freezers, and cold boxes. The transportation, shipment, and distribution of shipment are organized and provide preserved goods at the right temperature from origin to consumer. A cold chain is basically a supply chain regulated by temperature from origin to the end-use or destination. Depending on the item, the temperature typically varies from 2 to 8 degrees Celsius. The common commodities are such as seafood, agricultural produce, frozen foods, chemical, photographic film, vaccines, and pharmaceutical drugs
5	Urban or city logistics	Are about private companies in urban regions, process of total optimization in their transport and logistics activities. Considering the environment of the traffic, energy consumption of a market economy, and congestion during the distribution of commodities. Provided precedence toward the benefits and cost factors to the private sector including the public sector. Both objectives are making improvements, the public sector looking into the traffic congestion improvement and environmental issues. Whereas the private companies are focusing on cost reduction. Some examples of urban or city logistics such as Grab Food, Food Panda, Dah Makan, and more
6	Contract logistics	The definition is about the comprehensive process of the manufacturing or production till delivery of goods to the destination of sale. It is not just a matter of movement of goods, however an action or merging a traditional logistics in the processes of supply chain management. The strategic partnership is established in this contract logistics for a specific term and condition

(continued)

(continued)

No	Type of logistics	Definition
7	Business logistics	Can be referred to entire processes or a cluster of associated that incurred in the moving of goods and information from the sources of raw materials through to final consumers until recycling or disposal. This is inclusive of raw materials, movement, storage, final consumers, recycling, or disposal
8	Global logistics	Is defined as a project and administration of system in controlling the frontward and reverse movement of goods, information, and services ingoing and outgoing of the international corporation or across national boundaries from producers to consumer. Global logistics involves the movement of goods by truck, train, ship, or plane as well as preparation, packaging, and storage of goods in distribution centers and other logistics real estate facilities [10]
9	Event logistics	Is about integrated planning, managing, storing, and additionally tangible and insubstantial processes accomplished in an occasion. Organization takes account of aspects of program and place of the event. Potential modification in the general plan should make sure to run even more smoothly
10	Green logistics	Or sustainable logistics are the efforts on minimizing and measuring the impact on environment through logistics activities. Looking into proactive of the logistics activities process and operation of an organization to avoid damages to the environment. Its primary objective is to coordinate logistics activities within the supply chain for valuable and harmless to environment. This is associated with air pollution, climate change, noise, soil degradation, accidents, vibration, and dumping waste (including packaging waste)
11	Lean logistics	Which are through lean philosophy, means how one will design the manufacturing space specifically to minimize the integration of mankind, especially in a Covid-19 pandemic scenario. Looking into the overall logistics activities for effective and efficient value chain of the business internally. In other words, an assurance and offering a conducive workplace. Furthermore, improvement in occupations, employee well-being, improves performance inside the organizations
12	Reverse logistics	The role of logistics activities on focusing the backward movement of goods from end consumer back to customer is defined as the reverse logistics. In addition, a critical operation of the green SCM is for minimizing the waste commodities. The occurrence of product withdrawn may be resulted through numerous reasons from a various place along the supply chain including customer-related returns, distribution, and manufacturing. The predominant role of RL in disposition of commodities involves activities linked with decision on procedures and process of used commodities or returned commodities. Involving, remanufacturing, repair, reuse, disposal, and reprocessing [11]

(continued)

(continued)

No	Type of logistics	Definition
13	Smart logistics or logistics 4.0	Is the newest development in logistics practices which deal with technology of smart devices. Anticipation of smart technologies to bring about an increase in efficiency, safety, and business performance for companies [12]. The key factor in all logistics processes is the smart device, particularly in the field of cost reduction, performance enhancement, and market competitiveness. Approaching into the industry 4.0 (IR4.0) impact, potential logistics prospects, smart goggles, radio-frequency identification technology, smart gloves, and autonomous transport vehicles and self-transporting vehicles are reviewed [13]

6 Recent Trends of Logistics Study

Moving forward and concern toward sustainability coupled with digitalization in the supply chain management, the recent trends of logistics study possibility are insight on the smart logistics or logistics 4.0. Hence, the emergence of industry 4.0. provides opportunities for prospects in relation to new business models. Additionally, this studies looking into the importance of automated solutions, Internet of things (IoT), blockchain technology, instant information exchange, real-time big data, and other technologies enhancement.

7 Future Areas of Logistics Scholar

The functions and roles of information technology are adversely changing drastically in our daily experiences. Studying on the directive of women involvement toward economic stability. The issue addressing gender imbalance has buildt demand critically in the women talent in relation to logistics management practices. Further empirical research of the logistics study can be highlighted on the advantages of new technologies toward sustainability. On the hand, the need to foster logistician or non-logistician especially women is crucial to address gender inequality, for example, the Chartered Institute of Logistics and Transport, Women in Logistics and Transport collaboration with universities and industries to foster women knowledge and career advancement.

8   Conclusion

Previous research had shown limited study on logistics management and practices definition for logisticians. Further research on this topic area should develop a superior perception for logisticians including non-logisticians. Additional insight of logistics management and practices is to be analyzed to accomplish global consensus. Therefore, the logistics management and practices definition could generate interest for a holistic education, specifically for women to broaden their existence in this male dominance of logistics management and practices. This chapter highlighted logistics management and practices classification, recent trends, and future areas of logistics study for logisticians and non-logisticians.

Acknowledgements   This study is supported by Universiti Kuala Lumpur.

References

1. Z.A. Majid, M.F. Shamsudin, N.A. Rahman, *Halal Supply Chain Integrity Concept, Constituents, and Consequences*, 1st edn. (2023). https://www.routledge.com/Halal-Supply-Chain-Integrity-Concept-Constituents-and-Consequences/Majid-Shamsudin-Rahman/p/book/9781032305561
2. M. Christopher, *Logistics and Supply Chain Management, Financial Times* (2016). https://www.martin-christopher.info/logistics-and-supply-chain-management-5th-edition
3. Council of Supply Chain Management Professional (2011). https://cscmp.org/CSCMP/Educate/SCM_Definitions_and_Glossary_of_Terms.aspx
4. OECD, *OECD Competition Assessment Reviews: Logistics Sector in Malaysia* (2021). https://www.oecd.org/competition/fostering-competition-in-asean.htm; https://www.oecd.org/countries/bruneidarussalam/fostering-competition-in-asean.htm
5. Z.A. Majid, M.F. Shamsudin, N.A. Rahman et al., Advances in transportation and logistics: innovation in logistics from 1PL toward 10PL: counting the numbers. Research **2**, 440–447 (2019)
6. D. Topolseki et al., Defining transport logistics: a literature review and practitioner opinion based approach. Transport **33**(5), 1196–1203 (2018). https://doi.org/10.3846/transport.2018.6965
7. Z.A. Majid et al., Halal integrity from logistics service provider perspective. Int J Sup Chain Mgt **8**(5), 2444 (2019)
8. N.A.A. Rahman, Z.A. Majid, M.F.N. Mohammad, M.F. Ahmad, S.A. Rahim, The Development of Halal Logistics Standards in South-East Asia: Halal Supply Chain Standards (MS2400) as a Principal Reference (2020), pp. 149–160. https://www.taylorfrancis.com/chapters/edit/https://doi.org/10.4324/9780429329227-12/development-halal-logistics-standards
9. Department of Standard Malaysia, MS2400-2010 Halal Supply Chain
10. T. Alfonso et al., Lean thinking, logistic and ergonomics: synergetic triad to prepare shop floor work systems to face pandemic situations. Int. J. Global Bus. Compet. **16**(Suppl 1), S62–S76 (2021). https://doi.org/10.1007/s42943-021-00037-5
11. S. Banihashemi et al., Exploring the relationship between reverse logistics and sustainability performance: a literature review. Mod. Supply Chain Res. Appl. **1**(1), 2–27 (2019). https://doi.org/10.1108/MSCRA-03-2019-0009/full/html

12. R. Jurenka, D. Cagáňová, N. Horňáková, The smart logistics, in *Mobility Internet of Things 2018. Mobility IoT 2018. EAI/Springer Innovations in Communication and Computing*, eds. by D. Cagáňová, N. Horňáková (Springer, Cham, 2020). https://doi.org/10.1007/978-3-030-30911-4_20

13. J.O. Strandhagen, L.R. Vallandingham, G. Fragapane et al., Logistics 4.0 and emerging sustainable business models. Adv. Manuf. **5**, 359–369 (2017). https://doi.org/10.1007/s40436-017-0198-1

Effective Warehouse Space and Layout Management: A Case Study of DHL Express

Jamilahtun Md. Ghazali, Nik Nuralia Husna Nik Rushdi, and Hairul Rizad Sapry

Abstract Warehouses are critical components of a manageable business in operation flow and inventory management. This study examined the courier company's perspective on warehouse management and applying their space utilization. Therefore, warehouse planners must deal with certain characteristics that constrain the available surface area. Warehouse space utilization is critical for sustaining a smooth operating flow. This research concentrated on the space utilization and arrangement of warehouse space at DHL Express Service Point Glenmarie in Shah Alam, Selangor. The issue that frequently occurs in the warehouses is inefficient operation flows due to the space and layout. Each space in the warehouse should be meticulously planned, considering the warehouse's size and location. DHL Express Service Point Glenmarie is known for courier services, and the parcels or commodities will remain at the warehouse indefinitely and may be in excess. This results in insufficient space, resulting in a packed warehouse and lacking utilizing places. This study ascertains how the corporation utilizes space and warehouses to avoid the scenarios above. There are two primary objectives to accomplish. The first objective is to identify the effectiveness of the DHL Express Service Point space and layout in Glenmarie, Shah Alam, Selangor, in terms of parcels or commodities management. The second objective is to examine the suitable warehouse layout for DHL Express Service Point operation in Glenmarie, Shah Alam, Selangor. In this case of DHL Express Service, the suitable type of warehouse has been identified is the I-shaped layout. This suitable design is based on the operation optimization of DHL warehouse, budget consideration, space availability, equipment, and operation flow.

Keywords Warehouse · Space utilization · Layout design

J. Md. Ghazali (✉) · N. N. H. N. Rushdi · H. R. Sapry
Industrial Logistics, Universiti Kuala Lumpur, Malaysian Institute of Industrial Technology,
81750 Masai, Johor, Malaysia
e-mail: jamilahtun@unikl.edu.my

N. N. H. N. Rushdi
e-mail: nuralia.rushdi@s.unikl.edu.my

H. R. Sapry
e-mail: hairulrizad@unikl.edu.my

1   Introduction

Warehouse space and layout have become critical components, particularly for courier service companies. There are numerous considerations and fundamental principles that must be incorporated into warehouse floor planning. These procedures are designed to create a warehouse floor plan that is both cost-effective and functional. Another purpose of this research is to determine the space usage of warehouses. One of the issues confronting businesses in the delivery service industry and key hurdles is inadequate space for parcels/goods handling in warehouses. When a warehouse reaches approximately 80% of its capacity, it is theoretically overcapacity. However, this is not limited to the storage area; it also applies to receiving, shipping, and other business areas. It takes meticulous planning to create the ideal warehouse space and layout design while also managing a profitable and efficient business. If done incorrectly, the organization will pay significant costs [1].

The management of space and layout of the warehouse involves the overall structure (layout, dimensioning, and size), operational strategy (storage-picking), evaluation for layout, and control of picking operations. Each capacity must be sorted and organized to receive parcels or commodities properly. There must be reserved locations for processing parcel orders, loading and unloading commodities, and receiving orders for parcels. Proper warehouse layout benefits other processes, such as space use. As a player in the e-commerce support industry, this economic expansion runs parallel with DHL Express's growth. Furthermore, the excessive parcel and improper arrangement, lack of conveyors, racking, roll cages, etc., in the DHL Express Service Point Glenmarie, Shah Alam, Selangor will affect the warehouse's movement in the warehouse. There will be dumping of goods/parcels on the floor and cause the space in the warehouse to look packed and inefficient. The methods of warehouse utilization used by the organization will be examined in this research and why they are necessary to establish a utilization plan. Hence, the warehouse process's current state must be appropriately mapped out before analyzing it and planning the improvement. The research can also be helpful in terms of gaining new insights into warehouse management. As commonly known, the warehouse is one of the essential processes in the business, and this analysis has addressed the warehouse spaces management. In addition, understanding the warehouse space utilization and layout should be focus by the businesses to maximize their profit and investment [2]. This paper aims to answer the following research questions:

(1) What is the improvement have been made in DHL Express Service Point's warehouse space and layout?
(2) What is the impact of warehouse layout and optimization on DHL cost reduction and service efficiency?

2 Methodology

This research employs various techniques to ensure the accuracy of this report and that the strategies are collecting more accurate, precise, and appropriate information. For the analysis methodology, the qualitative approach has been done. Qualitative data analysis differs from quantitative data analysis. The qualitative data includes words, meanings, images, and even movements. The qualitative method gets in depth the results of the effective warehouse space and layout. It is nearly impossible to derive absolute significance from such data frequently used in experimental testing. While there is a significant difference between data collection and data analysis in quantitative research, evaluation for qualitative research often begins as soon as the evidence is available. Four aspects of qualitative research have been adopted as per Miles and Huberman [3], the location of the research, the participants, the event, and the process.

Primary data are gathered directly from primary sources through interviews, surveys, and studies conducted by researchers. Interviews were done with the senior managers and operation supervisor of DHL warehouse that have well knowledge on DHL warehouse management. They have been working with DHL Express more than 10 years and have a vast experience in managing the operations of DHL Express. Then, the data manual coding has been conducted. Coding summarizes the environment or individuals and categories or themes for analysis. All the data had been transcribed; the coding process began. For each theme, facts from each category will be grouped and collected. Furthermore, a table-shape representation is much easier to comprehend. The proof column has a coding scheme that provides descriptive evidence to support the study themes. Secondary data obtained from journals and book were used to support the data and justify the results from the interview [4].

3 Results and Discussion

The collected data were analyzed using the theme of the questions asked of respondents during the online interview session. This was critical for the researcher to obtain the desired findings and evaluations from the respondents' replies. Table 1 summarizes the findings.

A summary or brief discussion related to each research theme is a review of the findings. The researcher figured out that the findings connected with the objectives proposed.

The DHL Express Company's parcels/goods control has been demonstrated through the warehouse layouts implemented in their warehouses. The company's online interview session described and demonstrated the factors above in evaluating the space and warehouse layout and its effectiveness. Several complicating factors must be considered when designing a warehouse space and layout, and each design choice can significantly impact warehouse performance outcomes [5]. As a

Table 1 Summary of findings

Theme	Sub-theme	Keywords of finding
Theme 1	Areas in warehouse	• Three distinct warehouses for its parcel orders/goods • A flurry of activities involving equipment and also employees
	Capacity issues	• Drive-in racking full system • Efficient at utilizing space
	Procedure for loading/ unloading in the warehouse	• Different types of goods require different tools and equipment • Security and adequate training
	Areas/part in increasing warehouse's performance	• No main focus on areas/part increasing warehouse's performance
	Usage of warehouse space	• Organize storage by aisles • Increases total capacity and enables to maximize warehouse space
	Run out of space in the warehouse	• Operate efficiently and maximize warehouse space utilization • Requires a combination of techniques and methods
	Expected percentage of space utilization	• Approximately 85% of its capacity • Quickly assess the warehouse's storage space utilization
	Preparation parcels/goods before shipped out	• Many procedures • Shipping to guarantee compliance with both export legislation and destination state or country regulatory obligations
Theme 2	Issues in expanding the existing facilities	• Process before deciding on the layout • Create a blueprint and map of the warehouse • Hire a warehouse design specialist • No shortcut
	Preserve flexibility in the placement of parcels/ goods	• Planning and managing an appropriate warehouse layout • Easy to locate the higher-volume items
	Control over the number of parcels/goods	• SYSPRO—to maximize warehouse space utilization • Utilized a complete racking system, with each racking system having its unique numbering and control system
	Innovative technologies	• More expensive • Easy to set up • Lasts a long time • Most effective and revolutionary methods of warehouse storage utilization
	Performance of the company	• Experiment with the equipment • Employees will be evaluate • Alter the arrangements • The layout of an activity

result, DHL Express Service Point Glenmarie followed all procedures and processes to obtain the most suitable warehouse layout design. When all factors are considered, the company achieves the optimal use of space and design for layout and the highest level of storage management effectiveness. Since there are numerous factors to consider when designing a warehouse space and layout, each with its own set of specifications, determining the most critical first step can be challenging. However, the company provides warehouses with the most adaptable space and layout.

The researcher identified all the methods used by the company to achieve a functional warehouse layout. However, the researcher stated that each business has its level of implementation, indicating whether the methods are sufficiently effective or ineffective for parcels/goods management. The realism of an organization's data must be improved, and space utilization can play a significant role in this process. Improved data quality results in more efficient space utilization actions, which results in significant cost savings. Improved data integrity on a broader scale can also result in more accurate research and conscious decision-making across the organization [6].

DHL Express identified that a successful warehouse must have effective and efficient handling of parcels/goods. Additionally, if the warehouse contains flammable goods that could cause a fire, the company must take extra precautions, such as placing a fire extinguisher in each warehouse corner. Additionally, the warehouse must conduct at least a monthly audit by the safety team to check for potentially dangerous causes. To summarize, the company must employ various strategies to maximize space utilization in their warehouses to the tune of 80 percent.

4 Conclusion

The warehouse at DHL Express should utilize prominent signage and markings to facilitate navigation. Without adequate planning, the warehouse activity is likely to become unpredictable and inefficient. So, all employees must understand where and when they should be at all times. Begin by erecting appropriate signage throughout the warehouse to assist users and maintenance personnel efficiently and safely in navigating the facility. By clearly marking and specifying assigned user and equipment routes, the movement of parcels/goods and people are greatly simplified, reducing potential safety risks and increasing efficiency.

The aisle that contains the inventory is frequently used to describe the floor area of the warehouse. If the aisle is not utilized properly, the business's operations may suffer. It is simple as it eliminates unused space from the process and reduces the number of aisles. As a result, the number of pallet racks can be increased, also increasing the capacity of the warehouse. Reducing the aisle width from 11 to 12 to 6 feet to up to 60% more storage space, depending on the size of the racks and the items they hold.

Other strategies for increasing space utilization can be integrated into warehouse management. DHL Express Company may wish to define distinct boundaries for each warehouse section. More giant warehouses have the advantages of being more adaptable in terms of region utilization, allocating additional storage to zones, and even acquiring some unused space; smaller warehouses may not have that luxury. When renovating or arranging the warehouse, it is critical to establish proper boundaries for each area to maximize space and avoid overcrowding.

To conclude, the primary goal of this study was to assess the impact of the company's warehouse layout preparation on parcels/goods management and whether the company uses approaches to achieve space utilization. The findings indicate that the DHL Express Company has put in much work to their warehouse capacity and gone through selecting the best layout. However, the organization needs to work on a few points to grow its warehouse management system in the future ultimately. Real-time information and high-quality data are becoming increasingly important as a result of the complexity and variety of customer orders [7].

IoT is a new generation of embedded ICT devices that are Internet-connected and collaborate to combine supply chain and logistical operations in a digital setting [8].

Acknowledgements Special thanks to DHL Express Service Point Glenmarie Shah Alam Malaysia for their contribution and cooperation toward this study.

References

1. B.M. René, M. De Koster, A.L. Johnson, D. Roy, Warehouse design and management. Int. J. Product. Res. **55**(21), 6327–6330 (2017). https://doi.org/10.1080/00207543.2017.1371856
2. M. Živičnjak, K. Rogić et al., Case-study analysis of warehouse process optimization. Transp. Res. Proced. **64**, 215–223 (2022)
3. M.B. Miles, A.M. Huberman, *Qualitative Data Analysis: A Sourcebook of New Methods* (Sage, Thousand Oaks, 1994)
4. J.W. Creswell, J.D. Creswell, *Research Design: Qualitative, Quantitative and Mixed Methods Approaches* (Sage, Los Angeles, 2018)
5. Q. Ren, Y. Ku, Y. Wang et al., Research on design and optimization of green warehouse system based on case analysis. J. Clean. Product. (2023). https://doi.org/10.1016/j.jclepro.2023.135998
6. L. Wang, A.A. Hamad, V. Sakthivel, IoT assisted machine learning model for warehouse management. J. Interconnect. Netw. **22**(Supp02), 2143005 (2022)
7. D. Ahmed, *Warehouse Space Utilization: Tactics That Can Be Used to Improve* (2019). https://www.scmdojo.com/warehouse-space-optimization/. Accessed 20 Dec 2022
8. D. Kumar, R.K.R. Singh, R. Mishra et al., Application of the internet of things for optimizing warehousing and logistics operations: a systematic literature review and research direction. Comput. Ind. Eng. (2022). https://doi.org/10.1016/j.cie.2022.108455

A Cooling Roof Automatic System to Raise Building Temperature Efficiency

Muhd Syahrul Eizlan Saham, Zuhanis Mansor, and Izanoordina Ahmad

Abstract By 2050, it has been predicted that the gases released by air conditioners will be responsible for 27% of all global warming. This paper develops a green technology-based automatic cooling roof system for building applications. An automated control system maintains the indoor temperature independently of exterior weather conditions in the building, mainly the house. The Atmega328P with 32 kB in-system programmable flash memory, 1 kB EEPROM and 2 kB of internal SRAM turns on a sprinkler system and a fan to cool the building when the interior temperature rises above a predetermined level. If the temperature falls below a predetermined threshold, a heater is activated. In the Malaysian climate, the cooling mechanism is projected to operate significantly more frequently than the heating element. This system is cheaper and uses less energy than an air conditioner as it does not require an electrical compressor. This project uses green technology, using water as its cooling source. Since most air conditioners use fluorocarbons as their refrigerants and can prevent global air pollution, the project's long-term development would greatly benefit from installing a fire alarm and a heating system. Results reveal that the system provides a straightforward, reasonably priced way to create a comfortable home environment while reducing energy use.

Keywords Automatic cooling system · Green technology · Sprinkler system · Fan · Heater

M. S. E. Saham · Z. Mansor (✉)
Advanced Telecommunication Technology Research Cluster, Communication Technology
Section, Universiti Kuala Lumpur British Malaysian Institute, Batu 8 Jalan Sungai Pusu, 53100
Gombak, Selangor, Malaysia
e-mail: zuhanis@unikl.edu.my

M. S. E. Saham
e-mail: msyahrul.saham@s.unikl.edu.my

I. Ahmad
Electronics Technology Section, Universiti Kuala Lumpur British Malaysian Institute, Batu 8
Jalan Sungai Pusu, 53100 Gombak, Selangor, Malaysia
e-mail: izanoordina@unikl.edu.my

© The Author(s), under exclusive license to Springer Nature Switzerland AG 2024   79
A. Ismail et al. (eds.), *Technological Frontiers and Sustainable Innovations*,
SpringerBriefs in Applied Sciences and Technology,
https://doi.org/10.1007/978-3-031-68751-8_10

1 Introduction

The weather can be unpredictable, even though people did not experience sweltering weather in Malaysia. Instantaneously, the temperature can change from high temperature to low. Therefore, it is beneficial if the house's interior temperature can be maintained at a reasonable level. The sun's radiation heats the house during the day, particularly the roof, which raises the interior temperature above the normal range of 24–32 °C. Even with the fan operating, the temperature can be uncomfortable for us. One option for cooling the house is air conditioning, but it is a high-cost appliance that will be expensive to purchase and operate. It also requires a lot of energy. The use of coolants in the air conditioning system has a harmful environmental impact [1].

Numerous devices have been developed to lower a home's interior temperature, but most are independent, stand-alone devices. A microcontroller was used to coordinate the operation of these devices to fix this issue [2]. The Arduino microcontroller is identified for the automatic cooling roof system as it is less expensive and simpler to programme than most other microcontrollers [3]. By sprinkling water on the roof surface and turning on the fan to help the temperature drop more quickly, the automatic cooling roof system was able to regulate the indoor temperatures [4]. The low cost and high heat capacity of water or a water-based coolant led to their selection as the heat transfer medium. Since it uses a lot less energy to lower the house's internal temperature than air conditioning systems do, the system is environmentally friendly. When the house's interior temperature exceeds the set point for the temperature to cool the house, the sprinkler system and the fan will turn on by using the Arduino Uno as the primary component for the system to be operated [5–8]. The amount of energy used by air conditioning systems is sensitive to changes in load, ambient conditions, etc.

The main goal of air conditioning is to make the room's occupants comfortable by cooling the air. The above systems' issue is that the function cannot be controlled automatically; instead, it must be manually turned on and off. Additionally, this system occasionally generates high electricity usage, raising the electricity bill. Emissions from the air conditioner may also factor in the increase in global temperatures. No matter what the outside temperature is, the system is also unable to change the room's temperature.

Based on the ideas and the limitations mentioned above, this paper aims to provide an effective and economical solution to maintain a comfortable home environment while reducing energy consumption. This study embarks on the following objectives.

   i. To develop a cooling system that aids in the reduction of production and living costs in both households and industries based on the ATmega328P microcontroller.

  ii. To develop an environmentally friendly cooling system using the evaporative method.

 iii. To analyse the performance of the temperature efficiency using temperature sensors.

2 Methodology

Figure 1 depicts the project block diagram. The work's primary input is a temperature sensor. The 20 × 4 LCDs display the temperature after the temperature sensor has read the surrounding air temperature. The Arduino Uno R3 processes the digital output signal from the analogue input. The heater, water pump and fan are the main products of this project. In principle, the sprinkler system is a water distribution system that allows water to flow to the roof. There are three primary settings and conditions for this system. Whether the temperature range is within the temperature reference depends on that. The first condition will apply if the temperature is higher than the temperature reference. The fan and the sprinkler system will turn on in this scenario. The water pump motor will start with the voltage output before being used to power the sprinkler system. A second condition, as seen in the flow chart in Fig. 2, is when the temperature is within the reference range. In this scenario, only the fan will turn on; the sprinkler system will remain off. In this scenario, only the fan will turn on; the sprinkler system will remain off. The temperature is lower than the reference temperature in the final design, which is the third condition. In this case, the heater will turn on while the fan and water pump deactivates.

The circuit diagrams for the LM35 temperature sensor four and the DHT22 temperature sensor are shown in Figs. 3 and 4, respectively. Electronics' core components are used in circuit operations. The Arduino board's analogue input pin A0 is connected to the LM35 sensor's output. The potentiometer will set the desired temperature for the reference temperature and is connected to the second analogue input A1. Digital pin 11 is connected to a single eight-speaker. The DHT22 temperature and humidity sensor use a similar circuit. It uses the same circuit as the Arduino board for its data and output pins. The Vcc pin (+) and GND pin (−) of the sensor

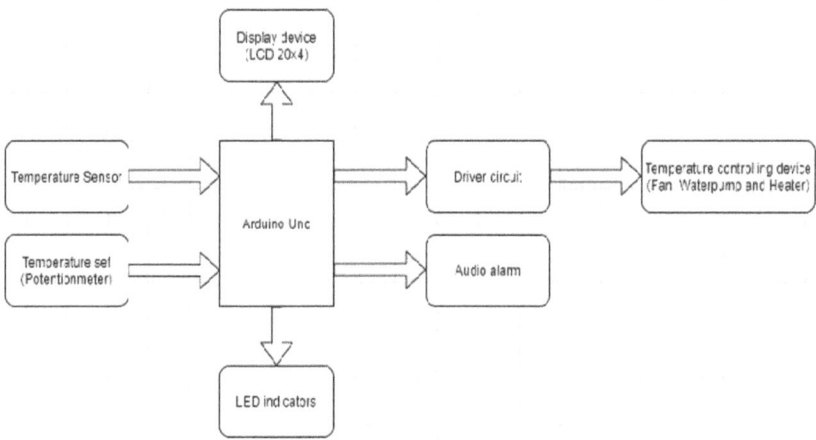

Fig. 1 Block diagram of the project prototype

Fig. 2 Flow chart of the
programming process

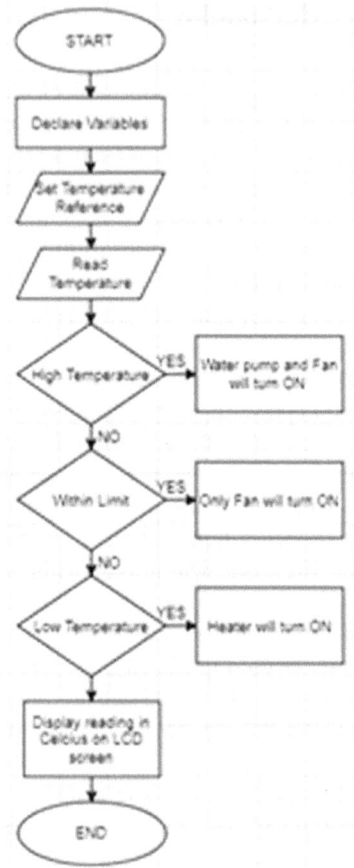

are wired to +5 V and ground, respectively. The prototype is tested on a scale model
of a house. The heating pad and temperature sensor are set up indoors. Water from
a water tank was pumped through tubing into the sprinkler system on the roof by a
water pump controlled by the water pump. For the house to receive the cool air, a fan
was installed underneath the roof. Figure 5 illustrates the rough drawing of the house
model. When water evaporates, a significant amount of heat is absorbed from the
environment. The most well-known instance is how perspiration cools human skin
through evaporation. The quick condensation of body heat from the skin's surface
regulates skin temperature in warmer, arid conditions. The high ambient air humidity
in hot zones with high humid air reduces the cooling effect.

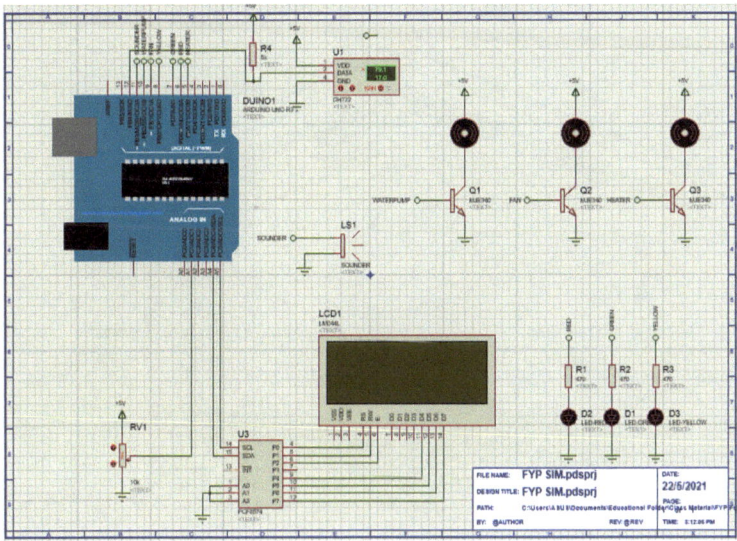

Fig. 3 The temperature sensor LM35

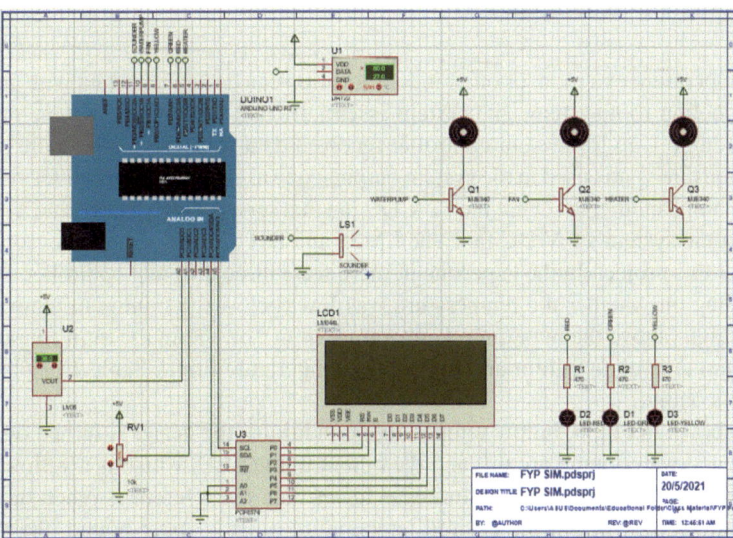

Fig. 4 Temperature sensor DHT22

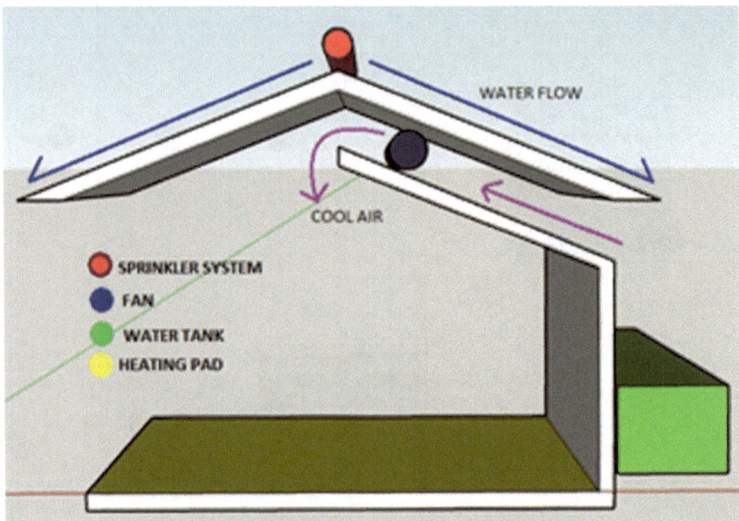

Fig. 5 House model for prototype's testing

3 Results and Discussion

The temperature of a room or other indoor location can be kept constant with an auto-matic cooling roof system. The system essentially had one input and three outputs. The Arduino Uno R3 system controller received data from the temperature sensor. There are three parts used in the system outputs: one to connect the sprinkler system's water pump, another to secure the 12 V DC fan that served as the heater, and a third to connect the system's 5 V DC fan. The LM35 and DHT22 temperature sensors were used for this project. In the experiment, there were three different controller settings depending on whether the temperature was low (below the minimum temperature limit), high (above the maximum temperature limit), or within the limit (in range of minimum temperature limit and maximum temperature limit). Table 1 displays the system's output.

As shown in Fig. 6, in low-temperature conditions, the heater is turned on while the fan and water pump are off, and the output to the LED, which is a yellow LED, was on when the room temperature was lower than the reference. When the room temperature falls within the desired temperature range's upper bound, the second

Table 1 Output parameters: fan, water pump and heater

Temperature (°C)	Output		
	Fan	Water pump	Heater
Low temperature	OFF	OFF	ON
Within limit	ON	OFF	OFF
High temperature	ON	ON	OFF

condition, i.e. within the limit condition, is met, and the green LED turns on. Only the fan will turn on in this scenario. As depicted in Fig. 7, the heater and the water pump were left in the off position. The output voltage turned on both the fan and the water pump at a high temperature in the third condition, where the temperature was higher than the reference temperature. The water pump powered up the sprinkler system, which was a water distribution piping system that directed water to the rooftop. Figure 8 depicts the high-temperature simulation.

Fig. 6 LM35 circuit—low temperature

Fig. 7 LM35 circuit—within limit

Fig. 8 LM35 circuit—high
temperature

4 Conclusion

The automatic cooling system described in this study provides a simple and cost-effective way to create a comfortable home environment while preserving energy. The system's circuit completed the goals and objectives of the design during the prototype trial run using a model house. Based on the concept of cooling after reaching a specific temperature in hardware, the design and construction of an automated temperature control system are developed in the Arduino-based hardware with a display. Results show that the system's circuit performed effectively in the prototype trial run utilising a model house that can cool the home by activating the sprinkler system and the fan when the house's internal temperature surpasses the set temperature.

Acknowledgements Zuhanis Mansor would like to thank the Advanced Telecommunication Technology (ATT) Research Cluster and Universiti Kuala Lumpur British Malaysia Institute (UniKL BMI) for the provision of laboratory facilities and financial support.

References

1. L. Joohyun, Y.L. Dae, Experimental study of a counter flow regenerative evaporative cooler with finned channels. Int. J. Heat Mass Transf. **65**, 173–179 (2013)
2. R.Y. Jeng, T.L. Juh, A microcomputer-based programmable temperature controller. IEEE Trans. Ins. Measur. **1**, 87–91 (1987)
3. Arduino—ArduinoBoardUno.   https://www.arduino.cc/en/main/arduinoboarduno.(2021). Accessed 20 May 2021
4. F. Li, H. Zhou, Y. Tao, Automatic temperature and humidity control system using air-conditioning, in *Paper Presented at the Transformer Substation. Asia-Pacific and Energetic Engineering Conference, Shanghai, China* (2012), pp. 27–29
5. A. Amoo, et al., Design and implementation of a room temperature control system: microcontroller-based, in *IEEE Student Conference on Research and Deviation* (2014)

6. M.Z. Adamu et al., Design and simulation of an automatic room heater control system. Heliyon **6**, 4531 (2018)
7. K. Sujatha, et al., Design and development of Android mobile-based bus tracking system, in *Paper Presented at 1st International Conference on Networks and Software Computing* (2014), pp. 231–235
8. J.S. Haberl, S. Cho, *Literature Review of Uncertainty of Analysis Methods (Cool Roofs) R to the Texas Commission on Environmental Qual* (Texas A&M University System, 2004)

Hazardous Cargo Practices Toward Green Port Performance: Evidence from Port of Tanjung Pelepas, Malaysia

Muhammad Muhsin Mohamad Ali, Amayrol Zakaria, Aminuddin Md. Arof, and Mohd Azam Din

Abstract The green port status is a novel idea for ports, particularly in handling dangerous goods in the maritime industry. Preventing and reducing pollution at the port is a critical factor to address in order to have the safest port operation. The term "green" refers to new ideas, innovations, and transitions, which includes the port's technical handling process. Because the green port is still in its initial stages of implementation, it is vital to determine what elements should be monitored by ports in their everyday operations. The aim of this study is to investigate the Port of Tanjung Pelepas current hazardous cargo practices and the relationship between hazardous cargo practices toward green port performance. Consequently, a purposive sampling method was used and 66 respondents from a total of 80 respondents were selected using the Krejcie and Morgan's table. On the other hand, the open-ended and closed-ended pilot tested questionnaire by relevant respondents is employed to obtain primary data, and the secondary data from the literature review have been translated into useful information by utilizing the qualitative content analysis. Next, descriptive analysis was used to determine the current hazardous cargo practices and the Pearson correlation was used to determine the relationship between the variables. The finding revealed that there is a strong relationship between the variables.

Keywords Green port · Dangerous goods · Hazardous cargo · Descriptive analysis · Pearson correlation · Response rate

M. M. M. Ali · A. Zakaria (✉) · A. Md. Arof · M. A. Din
University Kuala Lumpur Malaysian Institute of Marine Engineering Technology, Lumut, Perak, Malaysia
e-mail: amayrolzakaria@gmail.com

M. M. M. Ali
e-mail: muhsin.ali@s.unikl.edu.my

A. Md. Arof
e-mail: aminuddin@unikl.edu.my

M. A. Din
e-mail: azam@utar.edu.my

© The Author(s), under exclusive license to Springer Nature Switzerland AG 2024
A. Ismail et al. (eds.), *Technological Frontiers and Sustainable Innovations*,
SpringerBriefs in Applied Sciences and Technology,
https://doi.org/10.1007/978-3-031-68751-8_11

1   Introduction

Johor Port Authority (JPA) has developed the green port policy, which will serve as a general guide for all port operators in their efforts to achieve a green port environment. Under JPA's Strategic Plan 2013–2020, the green port policy has been selected as one of the key performance indicators (KPI) for establishing a safe and healthy port working environment. The author of [1] stated that port is a common term that denotes a conflict between human action and the environment, posing a threat to environmental protection. Many industrialized economies have unilaterally implemented green port policies and regulations in their nations as a response to sustainability and environmental challenges [2]. It was supported by [3] that developing ports without a good environmental and ecological preventive policy could endanger residents as well as nearby flora and fauna. The number of challenges or issues relating to the port's environment that have arisen in the maritime industry is now increasing. Various ports nowadays are attempting to achieve the "green ports" concepts or the "greener" status of port by introducing new technology and upgrading a system as well as ports infrastructure [4]. The main idea of "green port" development is the introduction of the terms "green" growth in the continued development of port systems and the implementation in environmental planning within port areas [1].

2   Literature Review

2.1   Hazardous Cargo

Hazardous cargo *(HC)* is loosely defined as dangerous cargo (DC)', hazardous materials (hazmat), or dangerous goods (DG) in maritime logistics literature [8]. Although these terms may be used interchangeably, the best way to distinguish between them is by highlighting the three functional areas in which dangerous substances are involved—production, logistics and transportation, and consumption [5]. The IMO documents make use of both dangerous goods and dangerous cargoes [6]. According to the United Nations (UN) classification, there are more than 3000 items listed as DG in the IMDG code. If mishandled, DG may result in accidents through toxic releases and explosions, thus impacting the environment [6].

2.2   Hazardous Cargo Handling Practices

The incorrect handling of HC during in-port transportation may result in collisions within the port waters, and accidents within the port area [7]. The International Maritime Dangerous Commodities Code specifies how various categories of goods, packages, or containers should be separated [6]. The code, which is amended every two years also provides for the appropriate codes, symbols, and terms used in the securement and segregation of HC. These must be closely adhered to in order to avoid and/or limit avoidable dangers [8].

2.3 The Challenges of Green Port Development

Previous studies have outlined green initiatives that could help the port to attain sustainability. To begin with, there's no global standard for green port indexes or certification [9]. However, an Egyptian study has proposed a green port performance index (GPPI) which is based on the relative weights assigned to key green port performance indicators, can be used to assess the overall greenness of a port in any country in terms of environmental regulations [10]. The implementation of a standardized green port index, a green port deal necessitates effective organization and leadership, suitable legislation and regulations, innovations, and a management system for environmental energy efficiency and sustainable development are important for the realization of green port attainment [12].

The risks associated with hazardous cargo management in a port environment are related to various aspects of humans, the port environment, infrastructure and facilities, and issues with port authorities and governing bodies [12]. On the other hand, PTP faced issues with dangerous goods (DG) that landed from foreign vessels [13]. A total of 110 containers were discovered abandoned at the Port of Tanjung Pelepas and were laden with dangerous electric arc furnace dust (EAFD) [13, 14]. Therefore, further investigation is needed into the current hazardous cargo practices at PTP and the relationship between the hazardous cargo practices and the green port performance in achieving sustainability development in the port industry.

3 Methodology

A convenience random sampling method was used and 66 respondents from a total of 80 respondents were selected using the Krejcie and Morgan's table. Nevertheless, only 60 respondents, which is a permissible percentage under the response rate, gave feedback due to 6 respondents indicating that they had insignificant experience being involved in the survey. On the other hand, the secondary data from the literature review have been translated into useful information by utilizing qualitative content analysis. A set of 7 Likert scale questionnaire was used as a primary research method to enable data collection during the survey. These questionnaires are divided into two categories which are open-ended and close-ended question. A content analysis and coding process were used to analyze the secondary data where the data from the literature review, journal, and books have been grouped into 8 themes for the purpose of questionnaire development.

4 Results and Discussion

4.1 Reliability, Descriptive Analysis and Pearson Correlation Analysis

Table 1 shows that the result of the Cronbach's alpha for the pilot test that represents the reliability analysis of the study. The Cronbach's Alpha value is 0.918 which means the set of questionnaires are reliable to utilize for the gathering process. The reliability statistic is given in Table 1. Based on Table 2, the highest mean value is "stowage and separation of the dangerous goods" which is the mean value of 6.6 followed by segregation standards with the mean value of 6.5, guidance on emergency response procedures with the mean value of 6.4, dangerous goods documents practice with the mean value of 6.3, the classification of the dangerous goods with the mean value of 6.3, packing material for dangerous goods are tested with the mean value of 6.2, the written permission from the Johor Port Authority or PTP with the mean value of 6.1 and the last is the activity of labeling with the mean value of 6.0. As for the standard deviation, all variables are less than 1.0 which indicated that 100% of the judgments are within 1, 2, and 3 SD, making the judgments for all variables are consistent with the Chebyshev theorem. Obviously, since all the variables are under the 6–7-point Likert scale (agree to strongly agree), it can be indicated that all variables or practices are currently in practice in the organization. Nevertheless, there is no suggestion from the expert respondent to add any new variable that is suitable for the study.

Table 3 indicates the degree of correlation between two variables which are the dependent variable and independent variable where 8 hazardous cargo practices are

Table 1 Reliability statistics

Cronbach's Alpha	Cronbach's alpha based on standardized items	No. of items
0.914	0.918	20

Table 2 Descriptive analysis

Practices	N	Mean	S.D
1. The classification of the dangerous goods	8	6.3	0.852
2. The written permission from the Johor Port authority or PTP	8	6.0	0.952
3. Packing material for dangerous goods are tested	8	6.2	0.878
4. The activity of labeling	8	6.1	0.882
5. Dangerous goods documents practice	8	6.3	0.852
6. Segregation standards	8	6.6	0.490
7. Stowage and separation of the dangerous goods	8	6.5	0.479
8. Guidance on emergency response procedures	8	6.4	0.727

Table 3 Pearson correlation analysis

	1	2	3	4	5	6	7	8	9
1	1.00	462	0.418	0.480	0.462	0.476	0.425	0.438	0.462
2	0.476	1.00	0.615	0.000	−0.255	0.240	−0.496	−0.373	−0.373
3	0.418	0.615	1.00	0.000	−0.191	0.881	−0.372	−0.063	−0.063
4	0.480	0.000	0.000	1.00	0.302	0.000	−0.592	0.000	0.000
5	0.462	−0.255	−0.191	0.303	1.00	−0.132	−0.205	462	0.462
6	0.476	0.240	0.881	0.000	−0.133	1.00	−0.258	0.257	0.257
7	0.425	0.496	−0.372	−0.591	−0.204	−0.258	1.00	−0.300	−0.300
8	0.500	−0.172	−0.062	0.000	0.462	0.258	−0.300	1.00	0.500
9	0.462	−0.255	−0.191	0.303	0.302	−0.132	−0.205	462	1.00

the independent variable and green port performance is the dependent variable. The "stowage and separation of the dangerous goods" is the highest value compared to another variable which is 0.500 and the data is close to 1 where the result demonstrates that there is a significant relationship between "stowage and separation of the dangerous goods" and a successful green port performance. On the other hand, the result also indicated that a positive value for all hazardous cargo practices or independent variables that signify a significant relationship with the green port performance. Obviously, all the hazardous cargo practices that have been analyzed above play an important role in influencing the green port performance at PTP.

5 Discussion

In retrospect, there are eight hazardous cargo practices that have been revealed in this study that could be observed that contribute to the success of the green port performance at PTP. The 60 qualified responses that made a great contribution to the survey at PTP typically agreed and believed that the eight practices above play an important role in achieving the successful green port performance at PTP. Moreover, the finding also revealed that most of the expert's respondent prefer to strictly practice the eight hazardous cargo practices that have been revealed in this study as a guideline for the handling of a dangerous good at any similar port in Malaysia. Additionally, the successful green port performance also can be achieved since the current infrastructure and the practices that have been provided by Johor Port Authority is sufficient for the cargo handling purpose. It is because Johor Port Authority takes safety precautions toward infrastructure and facilities at PTP by ensuring that all the standard operating procedure (SOP) are strictly followed by all crews, and stevedores. More importantly, SOP of Johor Port Authority is competent in achieving the green port performance, while talented to ensure the safety and security of cargo and environment as a whole.

6 Conclusion

In conclusion, in order to ensure the successful green port performance at PTP, there are several factors that need to be taken into account such as by delivering the best hazardous cargo practices for the best service quality to improve the green port performance. Ultimately, the finding revealed that there is a significant relationship between hazardous cargo practices and the successful green port performance. Last but not least, the further research is necessity to explore from the other perspective rather than hazardous cargo practices. For instance, form regulatory and the technological point of view and the other important variable that could be contribute to the successful green port performance. It is hoped that the sustainability of a port operation can be achieved by considering the three pillars which are environment, economic, and societal.

Acknowledgements The author's gratitude for the involvement of the expert respondents at PTP, Malaysia. The gratitude also for the Universiti Kuala Lumpur, Malaysian Institute of Marine Engineering Technology for providing a favorable environment to conduct this research.

References

1. P. Badurina, M. Cukrov, Č Dundović, Contribution to the implementation of "Green Port" concept in Croatian seaports. Pomorstvo **31**(1), 10–17 (2017)
2. A.M. Arof, A. Zakaria, N.S.F.A. Rahman, Green port indicators: a review. Int. J. Adv. Eng. Res. Sci. **12**, 237–256 (2021)
3. T.C. Lirn, Y.C.J. Wu, Y.J. Chen, Green performance criteria for sustainable ports in Asia. AJSL **12**, 134 (2013). https://doi.org/10.1108/IJPDLM-04-2012-0134
4. R.H. Chiu, L.H. Lin, S.C. Ting, Evaluation of green port factors and performance: a fuzzy AHP analysis. Math. Probl. Eng. **32**, 2976 (2014). https://doi.org/10.1155/2014/802976
5. A. Mah, *Port Cities and Global Legacies: Urban Identity, Waterfront Work, and Radicalism* (Springer, 2014)
6. International Maritime Organization, *IMDG Code: International Maritime Dangerous Goods Code: Incorporating Amendment*. IMO (2014). https://scholar.google.com/scholar?hl=en&as_sdt=0%2C5&q=International+Maritime+Organization+%282014%29.+Imdg+Code%3A+International+Maritime+Dangerous+Goods+Code%3A+Incorporating+Amendment+37-14+%28Vol.+1%29+IMO.&btnG. Accessed 20 July 2021
7. H.S. Loh, V.V. Thai, Management of disruptions by seaports: preliminary findings. APJML **64**, 53 (2015). https://doi.org/10.1108/APJML-04-2014-0053
8. F. Saruchera, Determinants of effective high-risk cargo logistics at sea ports: a case study. IJSM **14**(1), 1–13 (2020)
9. H. Wang, Assessing energy efficiency of port operations in china: a case study on sustainable development of green ports. J. Soc. Sci. Res. **3**(05), 28 (2015)
10. S. Elzarka, S. Elgazzar, *Green Port Performance Index for Sustainable Ports in Egypt: A Fuzzy AHP Approach* (IFSPA, 2014). http://www.icms.polyu.edu.hk/Proceedings/Proceedings%20of%20IFSPA%202014.pdf. Accessed 20 July 2021
11. C.C. Chang, C.M. Wang, Evaluating the effects of green port policy: case study of Kaohsiung harbor in Taiwan. Transp. Res. D Trans. Environ. **17**(3), 185–189 (2012)

12. R.U. Khan, J. Yin, F.S. Mustafa, Accident and pollution risk assessment for hazardous cargo in a port environment. PLoS ONE **54**, 252732 (2021). https://doi.org/10.1371/journal.pone.025 2732
13. NST, *110 Toxic Waste Containers from Romania Left at PTP* (2020). https://www.astroawani. com/berita-malaysia/110-containers-toxic-waste-romania-abandoned-ptp-251892. Accessed 20 July 2021
14. PTP Media Hub, *The Port of Tanjung Pelepas: Malaysia's Technologically Advanced Container Terminal* (2020). https://www.ptp.com.my/. Accessed 28 July 2022

A Delphi Study to Identify Important Green Port Indicators for Dry Bulk Terminals in Perak

Aminuddin Md. Arof, Amayrol Zakaria, Ismila Che Ishak, Hikmah Affirin Shahrul Alfian, and Noorul Shaiful Fitri Abdul Rahman

Abstract As the volume of cargo handled at seaports increases, the issue of ensuring the long-term sustainability of seaports has become an important topic at the international level. As this is a new initiative in Malaysia, very limited studies on green port can be found in the open literature. Since ports or terminals can be divided into several specializations, it is imperative for the port authorities and operators to produce their own guidelines to enable them to identify the actions that are required to be taken to achieve a green port status. Hence, the objective of this research is to identify the important green port performance indicators for dry bulk terminals by using the Delphi technique, as handling dry bulk cargo is considered one of the most challenging activities for the surrounding environment. The significant output of this research is to produce a conceptual framework on the degree of importance of a set of green performance indicators for dry bulk terminals.

Keywords Delphi technique · Dry bulk terminal · Green port · Sustainable port

A. Md. Arof (✉) · A. Zakaria · I. C. Ishak · H. A. S. Alfian
Universiti Kuala Lumpur, Malaysian Institute of Marine Engineering Technology, 32200 Lumut, Perak, Malaysia
e-mail: aminuddin@unikl.edu.my

A. Zakaria
e-mail: amayrol@unikl.edu.my

I. C. Ishak
e-mail: ismila@unikl.edu.my

H. A. S. Alfian
e-mail: hikmah.shahrul02@s.unikl.edu.my

N. S. F. A. Rahman
Faculty of Business, Higher Colleges of Technology, Abu Dhabi, United Arab Emirates
e-mail: nsfitri2107@gmail.com

A. Ismail et al. (eds.), *Technological Frontiers and Sustainable Innovations*,
SpringerBriefs in Applied Sciences and Technology,
https://doi.org/10.1007/978-3-031-68751-8_12

1   Introduction

More than 80% of global trade is carried on board ships and handled through the world's seaports. Therefore, without an efficient seaport and shipping network, the continually increasing demand to move international trade would not be effectively met. As port traffic continues to grow, the question of how to ensure long-term sustainability of the port sector is becoming an important issue at the international level. Since 2007, many developed economies have taken unilateral actions to implement green port policies and legislation in their countries. In Malaysia, initial steps towards the development of a green port policy were only announced in December 2016 and are based on three elements namely environment, community engagement and sustainability.

As this is a new initiative in Malaysia, a preliminary study shows that only the Johor Port Authority had publicized their green port policy in 2014. The green policy and initiatives taken by other local ports are arguably still minimal. Since ports or terminals can be divided into several specializations such as container, dry bulk, liquid bulk, passenger and general cargo terminal, it is imperative for the port authorities and operators to produce their own guidelines to enable them to identify the actions that are required to be taken for achieving a green port status. Hence, the objective of this research is to identify the important green port performance indicators at dry bulk terminals, as handling dry bulk cargo is considered one of the most challenging activities for the surrounding environment.

2   Aim

The aim of this study is to develop a conceptual framework on the degree of importance of a set of green performance indicators for dry bulk terminals by using a case study on the dry bulk terminals in the state of Perak, Malaysia. It is hoped that this effort could assist the operators of dry bulk terminals and the relevant regulatory authorities to identify and focus on the green indicators that can provide higher impact towards the sustainability of their ports.

3   Green Port

A green port is defined as a product of the long-term strategy for sustainable and climate-friendly port infrastructure development [1]. Sustainability in terms of the green port concept consists of three key elements: ecological balance, port economic stability and social development [2]. The concept of green port advocates with the port to minimize or eliminate harms to the environment and improve the port's efficiency. All of these positive effects will eventually affect workers' health and social stability, besides an increase in economic development. For a port to achieve a green status, it has to adhere to the green port concept and measure the port's green performance. There are many negative impacts from port operations. These negative impacts come in various forms such as pollution and environmental degradation. As their operations are continuous, the ports are aware of the pollution that is emitted to the natural environment [3]. These negative effects can be mitigated by implementing the concept of green port. A green port uses systems and technologies that prevent environmental pollution and enable to eliminate the negative impact of port activities on the environment [2].

4 Scope of Study

The scope of this study is limited to four dry bulk terminals in the state of Perak involved in the handling of iron ore, coal, grain and limestone. It can be argued that the cargo variety can be generalized to represent other dry bulk terminals in Malaysia. However, the survey respondents were not limited to experts working with the terminals. Other experts such as port users, port contractors, related government agencies and researchers were also invited to enable the collection of opinions from multiple perspectives. The ports in Perak can be divided into Lumut Port that operates two terminals, i.e. Lumut Maritime Terminal (LMT) and Lekir Bulk Terminal (LBT), Vale Malaysia Minerals that specializes in the handling of iron ore, and a private terminal owned by Malayan Flour Mill that is used to handle grain cargo.

5 Methodology

This study has employed a mixed-method approach. Initially, it involved the procedure of gathering qualitative data to explore the phenomenon (i.e. requirements for green bulk terminals). The gathering of qualitative data was performed through the literature review process and the open-ended question in the round 1 Delphi survey. A subsequent Delphi round was conducted to enable the achievement of consensus among expert respondents and subsequently identify the importance of the selected determinants by using a seven-point Likert scale.

At the planning stage of this study, 40 experts were chosen to participate in the Delphi survey. Experts were selected from the employees of the four terminals under study, port users, staff of the regulating agencies and researchers that have performed research on port operations. In order to be considered as experts, respondents must serve at supervisory level or higher with at least 5 years of experience. Based on the guidelines drawn, 40 participants were invited to participate in this study. However, only 24 experts have participated in the two Delphi rounds. The breakdown of the expert respondents is as per Table 1.

Table 1 List of expert respondents

Organisation	Lumut port	Vale minerals	Malayan flour mill	Bulk shipping	Enforcement agency	Researcher	Total
Invited	10	7	6	3	10	4	40
Responded	7	5	3	3	4	2	24

All the initial determinants and their clusters shortlisted in the round 1 Delphi survey are forwarded to be re-assessed by the same respondents in round 2. Since controlled feedback was given to the respondents, the median score for each determinant was included and respondents were allowed to change the judgements if they wish to do so. Only determinants with a mean score of between 6 (important) and 7 (very important) have been selected as important determinants that will contribute to the achievement of green port status for dry bulk terminals. Through the quantitative analysis of the outcome of the round 2 Delphi survey, 27 key determinants that were grouped under 7 clusters have been shortlisted by the expert respondents from a total of 58 determinants that were grouped into 13 initial clusters (Fig. 1). The shortlisted list is as per Table 2.

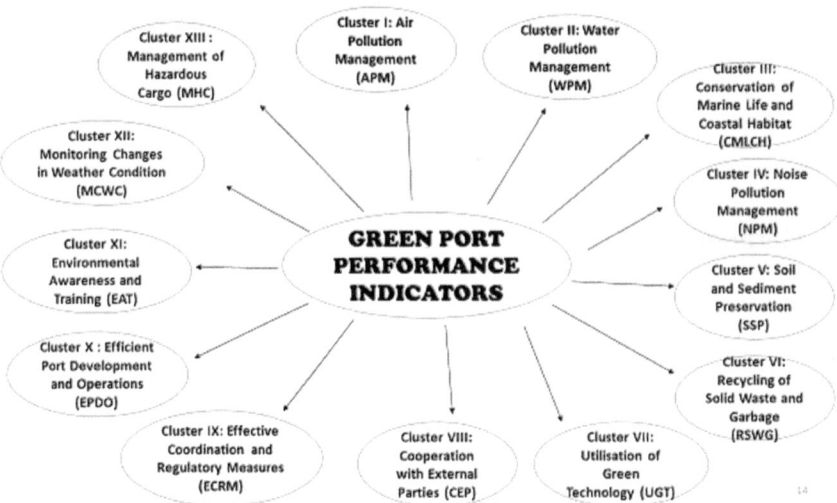

Fig. 1 Initial clusters for green port performance indicators for dry bulk terminals

Table 2 Determinants categorized as important by Delphi respondents

Cluster No	Cluster/determinant	Mean score
1	Air pollution management	
	a. Dust pollutant control	6.16
	b. Reduction of air pollution from toxic gas from ships	6.52
	c. Frequent dust removing and area cleaning	6.00
2	Water pollution management	
	a. Preventing ship bilge discharge	6.56
	b. Enforcement of oil spill control measures	6.80
	c. Sewage management	6.60
	d. Prevention of waste dumping	6.32
	e. Ballast water polluting control	6.56
3	Conservation and preservation of marine life, soil and sediment	
	a. Marine biology and wetland preservation	6.08
	b. Soil and sediment pollution prevention	6.00
	c. Strict control of dredging activities	6.04
	d. Monitoring and periodical analysis on activities that may affect port eco-system	6.17
4	Recycling of solid waste and garbage	
	a. Solid waste management	6.44
	b. Recycling of waste, garbage and resources	6.36
	c. Modernisation of waste management and facilities	6.12
5	Effective management, coordination, cooperation and regulatory measures	
	a. Adoption of ISO 14001 EMS	6.12
	b. Effective leadership	6.16
	c. Hazardous cargo management	6.48
	d. Collaboration to improve computer aided operations	6.08
	e. Standardisation of rules and regulations for all dry bulk terminals	6.17
6	Efficient port development and operations	
	a. Research on port development	6.16
	b. Efficient planning of port development	6.48
	c. Identification of key green projects	6.32
	d. Management of land transport and intermodal hinterland connection	6.00
7	Environmental awareness and training	
	a. Port staff training	6.28
	b. Development of awareness among workers	6.40
	c. Community awareness promotion and education	6.08

6   Results and Discussion

After the two rounds of modified Delphi survey, where the respondents were allowed to propose new determinants in the first round and were given controlled feedback on the aggregate results and allowed to change their judgement in the second round, only 27 out of the list of 49 determinants selected at the initial stage were considered as important to be implemented to enable dry bulk terminals to achieve a green port status. Due to the reduction in the determinants selected, the 27 important determinants have been re-arranged under 7 clusters as highlighted in Table 2 from the initial 13 clusters. Out of the 7 clusters, the most important cluster judged by respondents is the water pollution management (WPM) cluster, where the five determinants in the cluster obtained an average rating of 6.57 on a 7-point Likert scale. Besides the determinants under the preceding cluster, other determinants that were highly rated are 'reduction of air pollution from toxic gas from ships' with 6.52 point, 'hazardous cargo management' with 6.48 and 'efficient planning and port development' that also scores 6.48. The next three important determinants are 'solid waste management' that scores 6.44, 'development of awareness among workers' with 6.40 point and 'recycling of waste, garbage and resources' with 6.36 on a 7-point Likert scale.

When comparing the findings of this research with the top 10 environmental priorities given by European Ports in 2021, we can identify the similarities with six determinants chosen through the European Sea Port Organisation (ESPO) recent survey. Although the determinants have not been labelled with exactly the same title, they arguably cover the same activities. They are air quality, climate change, water quality, ship waste, port development (land related) and garbage/port waste [4]. The other top four determinants chosen by ESPO, i.e. energy efficiency, noise, relationship with the local community, and dredging operations were also chosen by the Delphi respondents as among the 27 important green port determinants [4]. These similarities arguably provide a construct validity to the findings of this research [5]. The differences in the priorities are arguably due to the focus of this research towards dry bulk terminals as compared to all types of ports in the ESPO survey and the slightly different priorities between developed Europe and developing Southeast Asia. This is proven from the changing of priorities from the nine ESPO surveys conducted from 1996 to 2021 [4].

7   Conclusion

In retrospect, this study was conducted to identify the important determinants that should be given priority by dry bulk terminals in their endeavour to be more environmental friendly. Priority needs to be given to the activities, which in the judgement of experts could bring greater impact on the green performance of the affected ports. Therefore, this study is not in any way trying to undermine the importance of the other determinants that were not shortlisted through the Delphi surveys. However, due to the limited resources experienced by many small port operators, this study could help in providing suitable guidance for the port operators or regulators to consider. Based on surveys conducted by ESPO, it can be observed that the green priorities of the European ports have changed over a period of 25 years. Therefore, continuous study on the important determinants for green dry bulk terminals should also be done in Malaysia and other parts of the world. Environmental priorities are expected to change as the size of the ports and ships becomes bigger, changes in cargo types and cargo handling equipment and not forgetting prevailing environmental issues experienced by the affected region.

Acknowledgements This work is supported by The Ministry of Higher Education Malaysia under the Fundamental Research Grant Scheme (FRGS), Grant Code: FRGS/1/2018/WAB05/UNIKL/02/1.

References

1. B. Pavlic, F. Cepak, B. Sucic, M. Peckaj, B. Kandus, Sustainable port infrastructure, practical implementation of the green port concept. Therm. Sci. **18**(3), 935–948 (2014)
2. Ł Marzantowicz, I. Dembińska, The reasons for the implementation of the concept of green port in sea ports of China. Log. Infrastruct. CEJSH **37**, 121–128 (2018)
3. P. Badurina, M. Cukrov, Č Dundović, Contribution to the implementation of "Green Port" concept in Croatian seaports. Pomorstvo **31**(1), 10–17 (2017)
4. ESPO Secretariat, *ESPO Environmental Report: EcoPortsinSights 2021, ESP-2844 (Sustainability Report 2021)* (2021). www.espo.be. Accessed 6 Sept 2022
5. S. Uma, B. Roger, *Research Methods for Business; A Skill-Building Approach* (Wiley, 2013), p. 227.

The Most Significant Factors on the Oil Spill Response Management among the Related Agencies in Malaysia

Ismila Che Ishak, Aminuddin Md. Arof, Md. Redzuan Zoolfakar, Mohd Fairoz Rozali, Hayatul Safrah Salleh, Ahmad Shahrul Nizam Isha, and Nur Aqilah Mohd Sabri

Abstract As far as we are concerned, once an oil spill incident occurs, it creates a critical long-term effect on the environment, society, economics, marine ecosystems, coastal communities, and human life. This paper analyzes the development of the significant factors of the oil spill faces by the related main government agencies in Malaysia such as the Marine Department and Department of Environment (DOE) together with the support from the key player private organization named Petroleum Industry of Malaysia Mutual Aid Group (PIMMAG), and from the other related

I. C. Ishak (✉) · A. Md. Arof · N. A. M. Sabri
Maritime Management Section, Universiti Kuala Lumpur, Malaysian Institute of Marine Engineering Technology, 32200 Lumut, Perak, Malaysia
e-mail: ismila@unikl.edu.my

A. Md. Arof
e-mail: aminuddin@unikl.edu.my

N. A. M. Sabri
e-mail: naqilah.sabri@s.unikl.edu.my

Md. R. Zoolfakar
Marine Engineering Technology Section, Universiti Kuala Lumpur, Malaysian Institute of Marine Engineering Technology, 32200 Lumut, Perak, Malaysia
e-mail: redzuan@unikl.edu.my

M. F. Rozali
Atase Maritim High Commission of Malaysia, 45-46 Belgrave Square, London SW1X8QT, United Kingdom
e-mail: fairoz@marine.gov.my

H. S. Salleh
Faculty of Business, Economics and Social Development, Universiti Malaysia Terengganu, 21030 Kuala Nerus, Terengganu, Malaysia
e-mail: hayatul@umt.edu.my

A. S. N. Isha
Management and Humanities Department, Universiti Teknologi Petronas, Persiaran UTP, 32610 Seri Iskandar, Lumut, Perak, Malaysia
e-mail: shahrul.nizam@utp.edu.my

© The Author(s), under exclusive license to Springer Nature Switzerland AG 2024   105
A. Ismail et al. (eds.), *Technological Frontiers and Sustainable Innovations*,
SpringerBriefs in Applied Sciences and Technology,
https://doi.org/10.1007/978-3-031-68751-8_13

agencies. It is essential to examine the significant factors of the oil spill response since Malaysia has faced 131 oil spill incidents from 2014 until February 2023 as reported by the DOE. This study has analyzed seven main significant factors that provide the most significant factors toward the oil spill incidents. The analysis was done on 59 expert responses received from 11 related agencies' concerns in responding to the oil spill incidents in Malaysia. Based on the factor analysis, the findings show that the suggested seven significant influences could be separated into three classifications which can be renamed and grouped as efficient leadership and response management, equipment management, and situational and effective communication. Meanwhile, the most important factors could be determined by the efficient situational analysis at 0.889, good leadership practices at 0.866, and coordinated decision making at 0.785.

Keywords Significant factors · Oil spill preparedness and response · Response agencies · Primary agencies · Contingency plan

1   Introduction

The oil spills are generated by numerous causes such as accidental leaks from ships and offshore oil platforms and these causes often rise in exorbitant economic costs and devastating marine ecological degradation [1]. The oil spill from tankers or other ships, which contain the transportation of liquid or bulk such as crude oil, fuel oil, or heating oil is a significant source of hydrocarbon inputs into the oceans, lakes, and rivers [2]. Once the oil spill incident appears, it will produce a negative impact on the marine environment as well as the maritime community at huge. The outcomes of the oil spill incidents on the environment and society can be tremendous. The suffocating and toxic effects of oil on flora and fauna could provide serious costs such as effects on public health, the local economy through loss of tourism and aquaculture, and historically or culturally significant sites [3]. Once, the pollution has saturated the water, it could generate destruction to marine life, caused marine life to die, disturb human health, and distract the ecosystems either on the sea or land [4]. The oil spill is anticipated to pollute a much greater area than primarily projected due to destructive weather conditions [5]. Even though an oil spill incident is occasional, once it transpires, it could lead to a tragic incident and damage to the sea, environment, people, offshore activities, fisheries, and economics, and crop a negative image of the country. Malaysia is highlighting the responsibility of pollution control, particularly in coastal and marine environments as it is crucial to meet water quality objectives. Malaysia is also committed to meeting the requirements of a quality environment as mandatory by the Environmental Quality Act 1974 [6]. Thus, this research aims to conclude the most significant factors toward the oil spill response in Malaysia gathered from 11 expert agencies who are dealing with the oil spill response in Malaysia.

Table 1 Oil spill cases in Malaysian for 2014–2020 [7, 8]

Year	No. of cases
February 2023	4
2021	7
2020	16
2019	13
2018	17
2017	22
2016	20
2015	15
2014	16
Total	130

1.1 Data on the Oil Spill Incidents

The data on the oil spill incidents in Malaysian water as presented in Table 1 were gained from the DOE from 2014 to February 2023 [7]. It shows that 130 oil spill incidents have been proclaimed as tier 1 or tier 2 [7]. As of now in May 2022, there are no oil spill incidents cases reported as a tier 3 disaster.

2 Literature Review

2.1 Challenges in the Oil Spill Response in Malaysia

The core difficulties challenged by the DOE in the oil spill issue are the nonattendance of manpower or staff to cover its enforcement, including the frequent changes and employment of a new officer after Service Circular No 3, 2004, which has obligatory an officer to enlarge the scope of work by increasing the productivity and deterring himself involves in any illegal tasks such as bribery, fraud, abuse of power, and misappropriation of money [9]. As the MMEA establishment is to hold out all federal deeds in the water, therefore the DOE faces a logistic challenge as the organization does not have any assets at sea and must cooperate with MMEA for assistance in resolving the issue associated with the oil spill incidents. On top of that, based on the reports on oil spill incidents at sea, most of the cases are from unknown sources and caused by illegal activities of ships along the Straits of Malacca and Singapore. Besides, an inadequate stockpile of oil spill equipment significantly contributes to a major problem faced by the DOE in the process of cleaning-up activities [7]. Besides, there is one major key industrial player response team from a private organization which is known as the Petroleum Industry of Malaysia Mutual Aid Group (PIMMAG). The organization was formed in December 1993 and up to May 2022, it has 29 registered members from oil and gas companies in Malaysia. PIMMAG will respond to their registered members in the stipulated time frame between 24 h to act at tier 1 of the oil spill incidents after being appointed by the ship-owner, agent, or terminal [9–12]. Ideally, once the oil spill has occurred, the response members from numerous disciplines are interdependent to some extent. The ad-hoc team is made accountable for the emergency response to solve the oil spill incidents immediately [10–12].

2.2 *Malaysia Oil Spill Contingency Plan (MoSCOP) 2021*

According to the DOE edition [13], has endorsed and approved Instruction 20 of the National Safety Council in March 2012 in which the oil spill incident is regarded as one of the maritime disasters. Therefore, under Article 7 Instruction 20, the National Safety Council and Department of Environment have been assigned as authorized agents using the National Contingency Plan of Oil Spill, while the Marine Department has been allocated as an executive agent for maritime disasters resulting from the oil spill incident or accident led to oil pollution or maritime disaster. The Malaysian National Oil Spill Contingency Plan (NOSCP) has been developed to control oil spill incidents occurring within Malaysian waters or waters comprised along Malaysian shorelines and the EEZ located within the Straits of Malacca and the South China Sea, as well as Brunei Bay, Sabah, and Sarawak waters and the Celebes Sea [11]. The NOSCP 2014 has been rebranded to Malaysia Oil Spill Contingency Plan (MoSCOP) [8]. The previous plan was recognized as Malaysian National Oil Spill Contingency Plan (NOSCP) [13].

2.3 *The Tier Response*

The National Contingency Plan has been stimulated by describing the three types of tiers which include tier 1, tier 2, and tier 3 based on the factors such as the locality of the spill, quantity of spill, and the capability and competency to respond. The oil risk and response are classified as corresponding to the size of the spill and its closeness to a company's operating capability. Table 2 shows the tiered reaction with the extent of the spill. Usually, tier 1 refers to a small spill, tier 2 refers to a moderate spill, and tier 3 refers to a huge spill.

Table 2 Tier response [14]

Amount of spilled oil	Proximity in operations		
Large			Tier three
Medium		Tier two	
Small	Tier one		

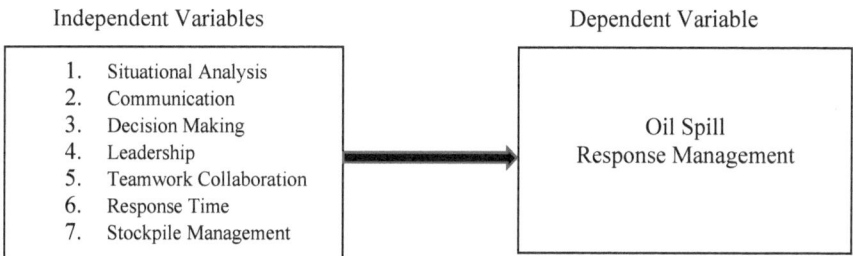

Fig. 1 Proposed summary of theoretical framework

3 Methodology

3.1 The Independent and Dependent Variables

The recommended theoretical framework in this study as shown in Fig. 1 and contains independent and dependent variables as the two key variables involved in social science research. The independent variable is the one the researcher controls and the dependent variable is the variable that varies in reaction to the independent variable. These two variables might be related to cause and effect. If the independent variable changes, then the dependent variable is concerned [15]. The independent variables involved in this research are situational analysis, communication, decision making, leadership, teamwork collaboration, response time, and stockpile management. Meanwhile, the dependent variable of this research is the oil spill response which requires regulatory, physical infrastructure, organization collaborations, and environment. It is based on the response elements modified and revised by the earlier findings [14, 16–18].

3.2 The Questionnaire Survey

The questionnaire survey is utilized in this study. It has been established to expedite the experts to ascertain the relative importance of the success factors concerned through a rank of the elements. It comprised of two sections, Section A: respondents' background and Section B: the determining significant factors in oil spill response. The questionnaire has been pilot tested by associates that have fundamental practice in marine pollution [19].

Table 3 Likert scale score of 7 points

1	2	3	4	5	6	7
Least important	Less important	Slightly important	Moderate important	Important	Very important	Most important

Table 4 Ratings of 14 successful factors in oil spill response

Factors	F1	F2	F3	F4	F5	F6	F7
Mean	6.34	6.51	6.54	6.51	6.51	6.29	6.37

3.3 Purposive Sampling of Respondents

Respondents who are identified as experts and have been involved in the oil spill response have been invited to participate in this survey. These respondents have considered the judgments of experts toward the oil spill to expediently priority scales. Authors of [20] suggest that the respondents should be restricted to those who have immediate knowledge and experiences. These targeted purposive samples cover those who have dealt with and experienced oil spill incidents and who have been subjected to oil spill operations and activities for at least three years. These response members have been visible and involved in a series of training and simulations linked to oil spill activities. The purposeful sampling method is widely used in qualitative research for the identification and selection of information gained in the research. This method is rich, and the cases are linked to the real occurrence of interest to the researcher [21].

3.4 The Pilot Test

A pilot test includes a small study to test research protocols, data collection instruments, sample recruitment strategies, and other techniques in preparation for research studies [22]. The pilot test entails the distribution of a questionnaire to five samples of the agencies and government authorities which comprised two from the Marine Department, two from the Department of Environment, and one from the Maritime Academy, respectively.

3.5 Likert Scale

A Likert scale is applied as an essential tool in psychology, social survey, and collecting attitudinal data. The Likert scales insist respondents choose among the 7-point level of agreement for that statement based on their point of view as given in Table 5. To determine the respondent's level of agreement with the statement

Table 5 Mean score interpretation [25]

Mean score	Interpretation
1.00–1.80	Very low
1.81–2.60	Low
2.61–3.20	Medium
3.21–4.20	High
4.21–5.00	Very high

presented in the questionnaires using an application of the Likert scale in an appropriate method [23]. The scale is commonly utilized in surveys or questionnaires, for benchmarking feedback in several fields [24].

4 Results and Discussion

4.1 Respondents Background

For the respondent's background, it reveals that most of the respondents were male, i.e., 45 respondents or 76.3%. This is because the prevalent gender functions in the maritime field are controlled by a male as this field demands masculine employees. The age of 31–40 years old is most of the respondents or with 32 respondents or 54.2%. This is because at this age most of the respondents have achieved their education study by having degree qualifications from various backgrounds to enable them to work in the marine-related field. It shows that most of the respondents have education at a degree level, i.e., 34 respondents or 57.6%. Meanwhile, for the current position, most of the respondents were in the middle management, i.e., 25 respondents or 42.4%. Thus, having a position in middle management inspires the employees to effectively have an excellent quality working attitude. Meanwhile, the experience in the oil spill below 3 years or 27 respondents or 45.8% contributes to the highest responses. This is because not all the respondents are subjected to oil spill incidents directly. Finally, the familiarity of the respondent with the oil spill incidents shows at a moderate level of 33 respondents or 55.9%. Hence, it is anticipated that this respondent profile provides integrity and credibility to the information collected over the questionnaire evaluation circulation.

4.2 Quantitative Analysis of the Significant Important Influences

To attain the objective of this study, the expert respondents were then expected to rate the importance of each factor on a seven-point Likert scale, beginning with 1 for

Table 6 Cronbach alpha interpretation [25]

Cronbach alpha	Internal consistency
$\alpha \geq 0.9$	Excellent
$0.9 > \alpha \geq 0.8$	Good
$0.8 > \alpha \geq 0.7$	Acceptable
$0.7 > \alpha \geq 0.6$	Questionable
$0.6 > \alpha \geq 0.5$	Poor
$0.5 > \alpha$	Unacceptable

Table 7 Reliability statistics

Cronbach's Alpha	No. of items
0.837	7

"Least Important" to 7 for "Most Important." All the 59 respondents that presented their feedback have been evaluated by the Statistical Package Social Science (SPSS) version 26 using mean analysis, total variance explained, factor analysis (FA), and Cronbach's alpha and frequency.

4.3 Mean Analysis of the Most Significant Factors

The mean score interpretation applied in this research is referred to by [25], as displayed in Table 5. Meanwhile, the standard deviation is also utilized to calculate how scattered the data is about the mean. A low standard deviation means data are gathered around the mean, and a high standard deviation indicates data are more spread out [26, 27]. In addition, Cronbach's coefficient of alpha internal consistency reliability is also utilized to evaluate the reliability of questionnaires as given in Table 6 [26, 27] (Table 7).

The internal consistency reliability is beneficial for understanding the magnitude to which the ratings from a group of judges are held simultaneously to determine a common dimension [26, 28]. Hence, a coefficient alpha of 0.837 from a maximum of 1 simply suggests that the scale scores obtained from the expert respondents are relatively reliable [26]. It shows that Cronbach's alpha reliability statistic of the most significant influences for the oil spill response is at 0.837 of the seven factors. Table 8 shows the average mean analysis for seven significant influences ranging from 6.29 to 6.54. Meanwhile, the average SD ranges from 0.626 to 0.751. All 59 items are valid and acceptable for data analysis.

Table 9 illustrates the most significant influences that could be categorized into three categories. *Group 1*: This factor analysis is utilized to establish the method of modeling the covariation among a set of observed variables as a function of one or more latent constructs [29]. This factor analysis has shown the rotated components matrix as the significant influence for the coordination efforts for the oil spill response

Table 8 Mean and standard deviation table of the most significant influences for the oil spill response

Factors	N	Mean	Std. deviation
Factor 1: situational analysis	59	6.34	0.734
Factor 2: communication among agencies	59	6.51	0.626
Factor 3: coordinated decision making	59	6.54	0.678
Factor 4: good leadership	59	6.51	0.751
Factor 5: good response time	59	6.51	0.653
Factor 6: stockpile management	59	6.29	0.696
Factor 7: harmonization administration	59	6.37	0.717

which can be grouped into three diverse groups as the rotated components matrix is high because more than 0.6 as given in Table 9. The new group for component one covers four influences which are factor 4, factor 3, factor 7, and factor 5. It is believed that good leadership and coordinated decision making among the response team could lead to better administration to provide a better service quality in the preparedness and response. Thus, the harmonization administration among the response team members could assist in having a good response time after hearing about the oil spill incidents. Thus, the new coding or name for the group one factor is classified as effective leadership and response management. *Group 2:* Meanwhile, component two covers only one factor of factor six on stockpile management which requires suitable stockpiles in response to the oil spill incidents. The effective logistics and management of the stockpiles reveal how to react after the oil spill incidents. Thus, the new coding or name for the group two factor is classified as equipment management. Finally, responding to oil spill incidents involves knowledge of situational analysis and communication among agencies. Thus, *Group 3:* consists of factors 1 and factor 2 which allow effective communication from all the response team members and may help to resolve the oil spill incidents effectively. The response team members expect to study and do the situational analysis of the incidents and to communicate effectively among all the response team members and other related agencies. Thus, the new coding or name for group three influences is classified as situational and communication.

Table 9 Rotated components matrix for the most significant factors of the oil spill response

Factors	Components		
	1	2	3
Factor 4: good leadership	**0.866**	0.282	−0.035
Factor 3: coordinated decision making	**0.785**	0.249	0.313
Factor 7: harmonization administration	**0.745**	0.172	0.339
Factor 5: good response time	**0.611**	0.564	0.176
Factor 6: stockpile management	0.131	**0.657**	0.523
Factor 1: situational analysis	0.139	0.259	**0.889**
Factor 2: communication among agencies	0.573	0.089	**0.708**

Extraction method: principal component analysis
Rotation method: Varimax with Kaiser normalization
[a]Rotation converged in 6 iterations

5 Conclusion

In retrospect, the questionnaire survey can be considered as successfully performed within less than two months the time taken for all the respondents to provide their responses. The researcher has performed an online questionnaire survey via a Google Form due to the Covid 2019 outbreak which allows the 59 respondents to react while the respondent is working from home and have adequate time to respond to this questionnaire survey. Subsequently, a few reminders were forwarded to the purposive sampling of the response. Based on the analysis, the results show that the suggested seven significant influences could be distributed into three categories which can be renamed and grouped as response administration management, equipment and experts support, and communication. Meanwhile, the most significant influences could be determined by the efficient situational analysis at 0.889, good leadership practices at 0.866, and large teamwork cooperation at 0.799. Thus, responding to the oil spill response requires the related agencies to have good response administration, handle efficient equipment, and continuous support from the related agencies is also crucial. Finally, communication among the related agencies must be effective too.

Acknowledgements This research is conducted and supported by the University Kuala Lumpur Malaysia Malaysian Institute of Marine Engineering Technology with cooperation from the UNIKL academic staff, supporting staff from external organizations together with other associated respondents from several agencies. Without good dedication and cooperation from the team members, it is hard for the research to be finished on time due to the Covid pandemic situation. This research financially obtained funding from the Ministry of Higher Education Malaysia through Fundamental Research Grant Scheme (FRGS) with grant number FRGS/1/2020/SSI03/UNIKL/02/1.

References

1. X. Shi, Y. Wang, M. Luo, C. Zhang, Assessing the feasibility of marine oil spill contingency plans from an information perspective. Saf. Sci. **112**, 38–47 (2019)
2. B. Doshi, E. Repo, J.P. Heiskanen, J.A. Sirvio, M. Sillanpaa, Effectiveness of N, O-carboxymethyl chitosan on destabilization of marine diesel, diesel and marine-2T oil for oil spill treatment. Carbohydr. Polym. **167**, 326–336 (2017)
3. B.L. Chilvers, G. Finlayson, D. Ashwell, S.I. Low, K.J. Morgan, H.E. Pearson, Is the way an oil spill response is reported in the media important for the final perception of the clean-up? Mar. Pollut. Bull. **104**, 257–261 (2016)
4. R.A. Anae, M. Alzuhairi, H.A. Abdullah, Corrosion behavior of steel (St 37–2) by using natural product as inhibitors in petroleum medium. Mater. Eng. **14**, 526 (2014)
5. S.Y. Chung, G. Lee, Combating oil spill accidents in Northeast Asia: a case of the NOWPAP and Hebei Spirit oil spill. Mar. Policy **72**, 14–20 (2016)
6. M. Mustafa, Environmental law in Malaysia. IIUM Law **19**(1), 1–34 (2019)
7. A. Norazaimah, *Statistic Oil Spill, Department of Environment (DOE) Putrajaya* (2019). http://www.doe.gov.my/. Accessed 22 June 2021
8. A. Azila. *Statistic oil Spill, Department of Environment (DOE) Putrajaya* (2021). https://www.doe.gov.my/portalv1/en/. Accessed 23 June 2021
9. R. Fairoz, *Perkongsian Maklumat Dengan Pihak Jabatan Alam Sekitar* (Jabatan Laut Malaysia, 2019). http://www.marine.gov.my/. Accessed 25 June 2021
10. H. Rahmat, PIMMAG: a joint industry effort in oil spill response and preparedness in Malaysia. Soc. Petrol. Eng. Int. **32**, 107–112 (1994)
11. C.K. Wing, Oil spill response management and transboundary issues in Malaysia, in *SPE Asia Pacific Health, Safety and Environment Conference and Exhibition. Society of Petroleum Engineers* (2005). https://onepetro.org/SPEAPHS/proceedings-abstract/05APHS/All-05APHS/SPE-96480-MS/89385 Accessed 23 June 2021
12. P. Li, Q. Cai, W. Lin, B. Cheng, B. Zhang, Offshore oil spill response practices and emerging challenges. Mar. Pollut. Bull. **110**(1), 6–27 (2016)
13. H. Hassan, *Rancangan Kontingensi Kebangsaan Kawalan Tumpahan Minyak (RKKKTM)* (Jabatan Alam Sekitar (JAS), Putrajaya, 2014)
14. I. IPIECA, *In-Situ Burning of Spilled Oil-Good Practice Guidelines for Incident Management and Emergency Response Personnel* (London, 2016)
15. T. Helmenstine, *Difference Between Independent and Dependent Variables* (2022). https://www.thoughtco.com/independent-and-dependent-variables-differences-606115. Accessed 20 Dec 2022
16. M.T. Crichton, K. Lauche, R. Flin, Incident command skills in the management of an oil industry drilling incident: a case study. J. Conting. Crisis Manag. **13**, 116–128 (2005)
17. R.S. Schuler, Definition and conceptualization of stress in organizations. Organ. Behav. Hum. Perform. **25**, 184–215 (1980)
18. G. Fink, *Stress Science, Neuroendocrinology*, 1st edn. (Academic Press, Massachusetts, 2017)
19. A. M. Arof, R. Nair, The identification of key success factors for interstate Ro-Ro short sea shipping in Brunei-Indonesia-Malaysia-Philippines: a Delphi approach. Int. J Shipping Transp. Logistics **9**(3), 261–279 (2017)
20. T.L. Saaty, How to make and justify a decision: the analytic hierarchy process (2002)
21. M.E. Reding, K. Guan, K.H. Tsai, A.S. Lau, L.A. Palinkas, B.F Chorpita, Evidence-based practice in child and adolescent mental health, **1**(2-3), 144–158 (2016)
22. Z.A. Hassan, P. Schatter, D. Mazza, Doing a pilot study: why is it essential? Acad. Family Phys. Malaysia **1**(2–3), 70–73 (2006)
23. R. Dittrich, B. Francis, R. Hatzinger, A. Katzenbeisser, A paired comparison approach for the analysis of sets of Likert-scale responses. Stat. Model. **7**(1), 3–28 (2007)
24. G. Pescaroli, O. Velazquez, A. Ayala, C. Galasso, P. Kostkova, D. Alexander, A Likert scale-based model for benchmarking operational capacity, organizational resilience, and disaster risk reduction. Disaster Risk Sci. **11**, 404–409 (2020)

25. K. Moidunny, The effectiveness of the national professional qualification for educational leaders (NPQEL). Doctoral Dissertation, Bangi. The National University of Malaysia (2009)
26. U. Sekaran, R. Bougie, *Research Methods for Business: A Skill Building Approach* (Willey, New York, 2016)
27. G. Ursachi, I.A. Horodnic, A. Zait, How reliable are measurement scales? External factors with indirect influence on reliability estimators. Proced. Econ. Financ. **20**, 679–686 (2015)
28. J.W. Osborne, *Best Practices in Quantitative Methods* (SAGE Publications, United States, 2008)
29. D.L. Bandalos, S.J. Finney, *The Reviewer's Guide to Quantitative Methods in the Social Sciences* (New York, 2018)

The Effect of Irresponsible Disposal of Face Masks to the Marine Ecosystem in Port Klang, Malaysia

Aizat Khairi, Qistiena Marsya Razali, and Shaiful Bakri Ismail

Abstract This research is to study the effects of irresponsible disposal of face masks and to propose the best way to dispose face masks. By using a questionnaire, data from the public in Port Klang has been collected and analyzed. The questionnaires collected the public demographics information, investigated the effect of irresponsible disposal of face masks and proposed the best way to dispose face masks. This study was aimed to suggest the best way to dispose face masks. A total of 396 respondents were collected and were analyzed using the Statistical Package for Social Science (SPSS). The findings are that the best way to dispose face masks is by using incinerators. It is important to know that face masks are contaminated with viruses, blood, or human body fluid. Therefore, the workers' health and safety must be put as priority in designing the bins.

Keywords Face masks · Marine ecosystem · Port Klang

A. Khairi (✉)
Faculty of Social Sciences & Humanities, Universiti Kebangsaan Malaysia, 43600 Bangi, Selangor, Malaysia
e-mail: zat@ukm.edu.my

Q. M. Razali
Maritime Management Section, Universiti Kuala Lumpur Malaysian Institute of Marine Engineering Technology, 32200 Lumut, Perak, Malaysia
e-mail: qistiena.razali@s.unikl.edu.my

S. B. Ismail
Marine Engineering and Electric Section, Universiti Kuala Lumpur Malaysian Institute of Marine Engineering Technology, 32200 Lumut, Perak, Malaysia
e-mail: sbakri@unikl.edu.my

A. Ismail et al. (eds.), *Technological Frontiers and Sustainable Innovations*,
SpringerBriefs in Applied Sciences and Technology,
https://doi.org/10.1007/978-3-031-68751-8_14

1 Introduction

The Covid-19 morbidity has a tremendous impact on human health, the economy, and daily life. Face masks are used as personal protective equipment (PPE) to minimize illness transmission. Waterlogged masks, gloves, hand sanitizer bottles, and other coronavirus misspend have been found on seafloors and beaches, adding to ocean waste [1]. Disposed face masks may generate chemical contaminants and nano-plastics. This increases plastics in aquatic medium. Disposed masks may not disintegrate for 400 years [2], and they harm animals. According to a study published in Animal Biology, this has happened to swans, seagulls, peregrine falcons, and songbirds, sometimes fatally [3]. Face masks generate millions of tons of plastic waste in a short time. Thus, the marine ecosystem longevity is at risk. The influence may also threaten human biology due to the food chain. This study is to identify whether worn face masks have degraded into marine waste, which could affect marine ecosystems. Oceans cover 70% of the earth's surface and are crucial to its health. Our abysses remain polluted [4]. Billions of pounds of trash and other contaminants penetrate our abysses. Face mass is generally made of non-renewable, petroleum-based polymers that are non-biodegradable, hazardous to the environment, and cause health problems. This issue may lead to environmental contamination, notably microplastic pollution in our abysses, and may harm marine ecosystems (e.g., seabirds, turtles, and crustaceans). This study aims to assist the public, marine life, and maritime industry in disposing of discarded face masks sustainably. Disposing of discarded face masks can prevent environmental pollution. It could also minimize the impact of face mask disposal on marine ecosystems by reducing mortality and increasing population.

2 Methodology

Quantitative research is the opposite of qualitative research, which involves non-numerical data (e.g., text, video, or audio). Quantitative research is used in biology, chemistry, psychology, economics, sociology, and marketing, among others [5]. Quantitative research develops knowledge and insight of the social world. Social scientists, especially communication scholars, use quantitative research to study human events. Quantitative research studies a sample population. Quantitative research uses scientific inquiry and measured data to analyze a sample population. A population is a group of people about whom data must be obtained. Sampling is the process of selecting a research group representative [6]. The population is the group from which the sample is drawn. A sample is the study or investigation group. Participants or respondents take part. As a result, the researcher chose a subset of the population that is representative of the target group and interested in the study of Port Klang which has roughly a population 32,655 [7]. The research has chosen 384 respondents to help with data collection.

A questionnaire consists of a sequence of questions meant to respondents. Questionnaires are a written interview. It can be done in person, over the phone, or on a computer. Questionnaires are an economical, rapid, and efficient approach to acquire significant volumes of information from a wide sample of people [8]. Because the researcher is not present when questionnaires are completed, data can be obtained fast. This strategy helps when interviews are impractical for big populations. These do not have predetermined answer selections and instead let respondents write what's on their mind. An open question is intended for difficult questions that need more information and discussion.

The pilot test is a trial run for the main study, allowing the researcher to test the research approach with a limited number of participants. Even though it is an extra step in the research process, it may be the best. The pilot test is required to criticize, test, and iteratively enhance the researcher's execution design [9]. This strategy ensures that the research proceeds smoothly and improves production. This study uses Cronbach's alpha to assess the internal consistency and reliability of survey or test items measuring the same construct. High Cronbach's alpha indicates internal consistency. A historical benchmark value of 0.7 indicates that some items measure the same concept [10]. The SPSS program was used to analyze the relationships proposed by the hypothesis supported by the theoretical framework [11]. The researcher applied descriptive statistics to present obtained data. The descriptive statistics are the gathering, organization, summary, and presentation of data. Descriptive statistics is utilized in this study to provide a short summary of the sample and measures. Charts, tables, and histograms are used to give quantitative information in an intelligible style [12]. This graphic analysis simplifies a lot of facts in a logical and appealing way.

3 Results and Discussion

The respondent's age is classified into six categories which is below 18–20 years old, which were about 56 respondents (14.1%), 21–30 years old which were 167 respondents (42.2%). The mean of Part A of the questionnaire is 2.81. This indicates that the maturity of the respondents is reflected in their responses to the questionnaire. The gender of the respondents gave a total mean of around 1.49. Most of these questionnaires were answered by the male (204 respondents), equivalent to 51.5%, and the rest was female, 191 respondents (48.2%) and 1 respondent (0.3%) preferred not to state the gender. This shows that the balance of responses is received from the questionnaire. For ethnicity, mostly Malays, with 231 respondents (58.3%) responded to this questionnaire. Since this questionnaire was distributed to the public in Port Klang, the respondents may come from various ethnic groups. Chinese ethnic was second-ranked, with about 74 respondents, bringing to 18.7%, then Indians with 45 respondents (11.4%). And lastly, 46 respondents from other ethnic groups answered this questionnaire survey—the mean for ethnicity was 1.76. This indicates that responses received were from various ethnic groups.

The level of education of respondents subdivided into five categories which is SPM/O-Level, with 57 respondents (14.4%), and level education of Diploma, which were about 110 respondents (27.8%) responded to this questionnaire. Most of these questionnaires were answered by a degree holder, i.e., 165 respondents. This shows that the level of education is comes along with their opinions toward certain topics. Types of employment, as examined, most of the respondents are students, with 152 respondents (38.4%) responding to this questionnaire. Since this questionnaire was distributed to the public in Port Klang, the respondent may come from several types of employment. The mean for types of employment is the highest mean with 2.95. This shows that the surveys cover various types of employment, therefore valid responses received.

Table 1 shows "strongly agree" is the answer to the question, how do you agree that irresponsible disposal of facemask induces harm to the marine ecosystem? This represents the mean of 4.2071. However, strongly disagree was the least answer from the respondent. This indicates that irresponsible disposal of facemask induces harm to the marine ecosystem. While Table 2 shows that 183 of respondents agreed that most of respondents are aware toward the irresponsible disposal of face masks in marine ecosystems. The mean result for this question is 4.1490. This shows that the existing waste already concerns the public and marine ecosystem.

Table 1 Induced harm

How do you agree that irresponsible disposal of face masks induces the marine ecosystem?

		Frequency	Percent	Valid percent	Cumulative percent
Valid	Strongly disagree	9	2.3	2.3	2.3
	Disagree	10	2.5	2.5	4.8
	Slightly disagree	40	10.1	10.1	14.9
	Agree	168	42.4	42.4	57.3
	Strongly agree	169	42.7	42.7	100.0
	Total	396	100.0	100.0	

Table 2 Marine species at risk

The existing and future situation of microplastics contamination, which revealed that all marine species are at risk of interacting with microplastics debris

		Frequency	Percent	Valid percent	Cumulative percent
Valid	Strongly disagree	6	1.5	1.5	1.5
	Disagree	12	3.0	3.0	4.5
	Slightly disagree	47	11.9	11.9	16.4
	Agree	183	46.2	46.2	62.6
	Strongly agree	148	37.4	37.4	100.0
	Total	396	100.0	100.0	

Table 3 shows that most of the respondents agree with the Covid-19 environmental impact, and this gives the mean is 4.0505. So, it can be summarized that the respondents, 173, are informed that Covid-19 activates an environmental impact. This indicates that the Covid-19 has had the impact of irresponsible disposal of face masks. For the issue of improper disposal of face masks, the public agrees and the mean is 4.0455. To summarize, the public respondents, with a frequency of 176 respondents, are sensible about the effect on the terrestrial environment. This shows that the disposed masks are the main issue in this situation. The response regarding disposable masks is plastic products that cannot be biodegraded but may be fragmented into tinier plastic particles; most respondents agree and know that face masks cannot be biodegraded. Most of the respondents agree that ingestion of these pollutants by organisms can result in asphyxiation (suffocation), serious injuries, inability to swim and dying from starvation. One hundred eighty-nine respondents decided to agree with this statement. This shows that these pollutants can result to asphyxiation to the marine ecosystem. The deposition of plastics can cause chronic biological impacts on the marine ecosystem. The mean for the seventh question is 4.0455. To sum up, most of the respondents agreed on these biological impacts.

On the other hand, most respondents agreed with this statement that waste management has a significant impact on people and the environment. Thus, mitigation strategies are the most significant potential to lessen the negative impact of improper face mask disposal in the marine ecosystem. Many people agreed with this statement in the questionnaire. Face masks must be disposed of according to specific guidelines which gives the mean of about 4.1566. 168 respondents (42.4%) of this questionnaire agreed with the statement, while 7 frequency (1.8%) strongly disagreed with the statement. Still, the need for specific guidelines to dispose masks is important to the people so that they will comply with it and reduce risks to the marine ecosystem. Table 4 shows the public awareness through social media platforms and regular programs and initiatives to support people's information. Most of the public agreed with this statement with the frequency of 189 respondents (46.8%), while 0 respondents strongly disagreed. It showed that the public wanted and had initiatives to gain their knowledge. This brings the mean to 4.1970. In terms of the encouragement to

Table 3 Unpredicted environmental implications

COVID-19 has had a number of unpredicted environmental implications, which included a reduction in recycling and a rise in the use of plastic globally

		Frequency	Percent	Valid percent	Cumulative percent
Valid	Strongly disagree	6	1.5	1.5	1.5
	Disagree	15	3.8	3.8	5.3
	Slightly disagree	67	16.9	16.9	22.2
	Agree	173	43.7	43.7	65.9
	Strongly agree	135	34.1	34.1	100.0
	Total	396	100.0	100.0	

explore possible approaches to developing biodegradable masks, most of the respondents answered the scale (4) agree with a frequency of 181 respondents and strongly agree with a frequency of 149 respondents. This item had given a mean of 4.1540. This indicates the need for possible approaches to develop biodegradable masks. Figure 1 shows the pie chart in regard to the classification of opinions on the best way to dispose face masks. Most of the respondents (34%) gave the opinion to burn it using incinerators as the best way to dispose. The second category is to segregate it using special bins. However, this is not the total solution to the used face masks.

Table 4 Public awareness

As a matter of fact, it is critical for countries to focus more on raising public awareness through social media platforms and implementing regular programme and initiatives

		Frequency	Percent	Valid percent	Cumulative percent
Valid	Disagree	11	2.8	2.8	2.8
	Slightly disagree	48	12.1	12.1	14.9
	Agree	189	47.7	47.7	62.6
	Strongly agree	148	37.4	37.4	100.0
	Total	396	100.0	100.0	

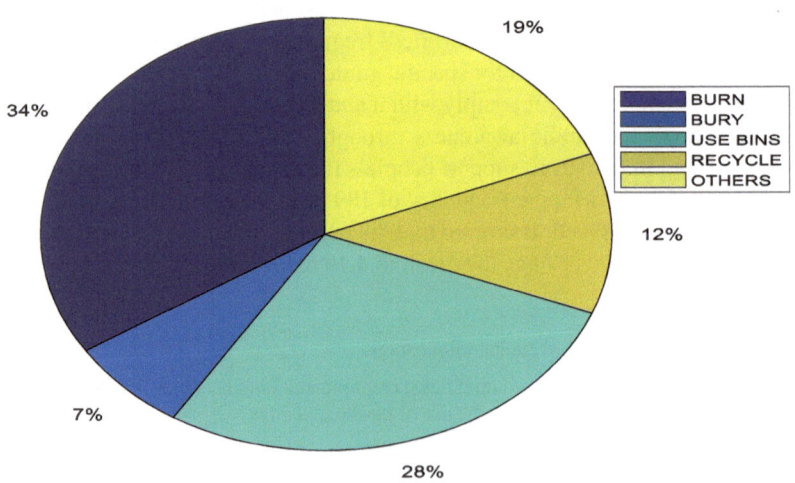

Classification of Opinions on the Best Ways to Dispose Face Masks

Fig. 1 Opinions on the best way to dispose face mask

4 Conclusion

This irresponsible disposal of used masks into the environment and mishandling of their debris could pollute marine ecosystems across the country. Slow breakdown of mask-derived polypropylene and polyethylene fibers produces microplastic contamination that affects marine species acutely and chronically [13]. When it comes to face masks, there must be a public use of essential personal protective equipment. Port Klang is a public research area; thus, respondents must have used face masks. The consequence after discarding the masks is obvious. Since the researcher observed that this irresponsible attitude threatens human health and the marine ecology, she decided to study the effect of face mask disposal on the marine ecosystem [14]. This study also proposes the best way to dispose face masks. Most public responses agreed with burning contaminated face masks in an incinerator. As of present, the incinerator is the best way to dispose of it without hurting marine ecosystems.

The following recommendations are based on the work accomplished during this study and on the conclusion given previously: The relevant ministry (Ministry of Environment and Water and Ministry Health of Malaysia) is to come up with policies. By creating policies on disposing face masks, it could mitigate the impact of face mask pollution in the future. The policies are to be consistently enforced by the authority so that the public will continuously comply with it thus achieving its target of reduction in pollution and collaborate with authority (Municipal Solid Wastes) and private companies that manage waste such as Alam Flora, E-Idaman, and so on. Collaboration with them could improve the level of public awareness through education at school, university, or even through social media. The school or university should demonstrate the value of taking care of the environment to the young generation so that they can live healthily and preserve it to the next generation. It is proposed to have dedicated bins to collect used face masks by collaborating with Klang Municipal Council. The bin should be suitable for its purpose, be made from suitable material and user-friendly for the public to use it. It also should be designed in such a manner that it will be easy to transport and not infect the workers that manage it. Remember that used face masks are contaminated with viruses, blood, or human body fluid therefore the workers' health and safety must be put as priority in designing the bin.

Acknowledgements We would like to express our thanks to Universiti Kuala Lumpur Malaysian Institute of Marine Engineering (UniKL MIMET) which provide us the funding to publish this paper and robust support in terms of research activity and innovation.

References

1. W. Ameya, B. Falguni, S. Raj, Plastic pollution. Res. J. Appl. Sci. Eng. Technol. **10**(3), 1564–1568 (2022)
2. A. Isobe, S. Iwasaki, The fate of missing ocean plastics: are they just a marine environmental problem? Sci. Total. Environ. **825**(153935), 1–13 (2022)

3. A.V. Boroda, Marine mammal cell cultures: to obtain, to apply, and to preserve. Mar. Environ. Res. **129**, 316–328 (2017)
4. S. Guo, Y. Zhang, H. Qu, M. Li, Repurposing face mask waste to construct floating photothermal evaporator for autonomous solar ocean farming. EcoMat **4**, 1–10 (2022)
5. T.C. Powell, Can quantitative research solve social problems? Pragmatism and the ethics of social research. J. Bus. Ethics **167**(1), 41–48 (2022)
6. H. Taherdoost, Sampling methods in research methodology; how to choose a sampling technique for research. How to choose a sampling technique for research. Int. J. Res. Manag. **5**(2), 18–27 (2016)
7. O. Merk, Shipping emissions in ports. Int. Transp. Forums Discuss. **12**, 83 (2014). https://doi.org/10.1787/5jrw1ktc83r1-en
8. K. Einola, M. Alvesson, Behind the numbers: questioning questionnaires. J. Manag. Inq. **30**(1), 102–114 (2014)
9. A. Gani, N. Imtiaz, M. Rathakrishnan, A pilot test for establishing validity and reliability of qualitative interview in the blended learning English proficiency course. J. Crit. Rev. **7**(05), 140–143 (2020)
10. M.A. Bujang, E.D. Omar, N.A. Baharum, A review on sample size determination for Cronbach's alpha test: a simple guide for researchers. Malays. J. Med. Sci. **25**(6), 85–100 (2018)
11. M. Yang, K. Suanpong, A. Ruangkanjanases, Development and validity test of social attachment multidimensional scale. Front. Psychol. **654**, 57777 (2022). https://doi.org/10.3389/fpsyg.2021.757777
12. B. Bozkurt, The relationship between social justice leadership and organizational citizenship behaviours. Particip. Educ. Res. **9**(2), 88–102 (2021)
13. H. Chowdhury, T. Chowdhury, S.M. Sait, Estimating marine plastic pollution from COVID-19 face masks in coastal regions. Mar. Pollut. Bull. **168**(112419), 1–7 (2021)

Evaluating the Performance of ResNet Variant Models for Car Model Detection Using a Transfer Learning Approach

Michael Chi Seng Tang and Huong Yong Ting

Abstract This paper presents a comprehensive investigation into car model detection utilizing state-of-the-art deep learning architectures, specifically focusing on popular ResNet variants: ResNet18, ResNet50, and ResNet101. A dataset consisting of 6730 car images across 33 distinct classes was meticulously curated. Transfer learning using ResNet101 and meticulous training procedures resulted in a highly accurate model achieving an accuracy of 83.40%, outperforming other models including ResNet18 and ResNet50. Comparative analysis of various methods underscores the superiority of the proposed ResNet101-based approach in accurately identifying car models. The study highlights the potential and relevance of advanced deep learning architectures, offering promising applications in real-world scenarios for precise car model recognition. Future research will continue to refine and extend these advancements to further enhance performance and broaden the scope of car model detection.

Keywords Car model detection · Deep learning · ResNet · Transfer learning · Image classification

1 Introduction

In the age of advancing technology and widespread utilization of image processing and computer vision, deep learning models have become extensively employed for various tasks, ranging from medical imaging [1–5] to car plate detection [6] and vehicle model recognition [7–10]. Recognizing car models not only holds importance

M. C. S. Tang (✉) · H. Y. Ting
Design and Technology Centre, University of Technology Sarawak, 1, Jalan University, 96000 Sibu, Sarawak, Malaysia
e-mail: michaeltang@uts.edu.my

H. Y. Ting
e-mail: alan.ting@uts.edu.my

© The Author(s), under exclusive license to Springer Nature Switzerland AG 2024   125
A. Ismail et al. (eds.), *Technological Frontiers and Sustainable Innovations*,
SpringerBriefs in Applied Sciences and Technology,
https://doi.org/10.1007/978-3-031-68751-8_15

in automotive security, traffic monitoring, and law enforcement but also finds applications in various domains like smart parking systems, insurance, and urban planning. This study delves into the realm of utilizing deep learning models, particularly ResNet variants, for precise car model detection.

The objective of this research is to explore and evaluate popular ResNet architectures, including ResNet18, ResNet50, and ResNet101, for their efficacy in car model detection. Prior studies have demonstrated remarkable success in utilizing deep learning, especially convolutional neural networks (CNNs), for image classification tasks. However, the specific focus on ResNet models, known for their deep architecture with residual connections, in the domain of car model recognition remains relatively unexplored.

The proposed investigation seeks to fill this gap by evaluating the performance of these ResNet models, assessing their accuracy, recall, and precision in identifying a diverse set of car models. Through transfer learning, where pre-trained models are adapted to suit the car model classification task, this study aims to discern the advantages and potential limitations of employing ResNet architectures in this context.

The ensuing sections will present related works in the domain of car model detection, discussing notable contributions and methods utilized. Subsequently, the methodology section will elucidate the data collection process, model training, and evaluation strategy. The results and discussions section will showcase the performance of ResNet18, ResNet50, and the proposed ResNet101-based model, providing a comparative analysis. Finally, the conclusions will summarize the findings and propose future directions for refining car model detection using deep learning methodologies.

2   Related Works

Several papers have been published that used image processing and computer vision techniques for car model detection. For example, Psyllos et al. [7] proposed an innovative vehicle manufacturer and model recognition (VMMR) scheme that leverages color recognition to enhance recognition accuracy. Utilizing a probabilistic neural network as a classifier and employing relatively simple image processing measurements, the system achieves robust vehicle authentication. It integrates license plate recognition, symmetry axis detection, and image phase congruency calculation, demonstrating high recognition rates and fast processing times suitable for real-time applications. While existing literature primarily addresses vehicle detection and general classification, VMMR, especially from frontal view images, remains a less explored area. The proposed system significantly contributes by effectively utilizing image processing techniques for license plate recognition, vehicle mask and logo image detection, and segmentation, while also employing a hierarchical image database to expedite recognition. Overall, the study presents a promising VMMR system that efficiently processes frontal view vehicle images for accurate and rapid recognition of both manufacturer and model, showcasing its potential for real-world applications.

Fang et al. [8] proposed a coarse-to-fine convolutional neural network architecture for fine-grained vehicle model recognition, addressing the challenge of subtle intra-category appearance variations. The approach involves detecting discriminative parts using feature maps generated by a CNN and establishing a mapping to locate these regions. The framework hierarchically extracts global and local features, which are then utilized in a one-versus-all support vector machine (SVM) classifier for classification. The CNN is trained using a pre-trained model initially, which significantly speeds up the training process. The experimental results demonstrate that the proposed framework outperforms state-of-the-art approaches, achieving an accuracy of 98.29% over 281 vehicle makes and models. The study contributes to advancing fine-grained vehicle model recognition through an intelligent and adaptable approach utilizing CNNs and SVMs.

Soon et al. [9] introduced a novel approach for vehicle model recognition using a principal component analysis network-based convolutional neural network (PCNN). Unlike existing methods that focus on global or multiple local features, PCNN specifically targets the vehicle headlamp as a discriminative local feature for recognition, eliminating the need for precise headlamp localization and segmentation. The PCNN model effectively combines principal component analysis and CNN to extract hierarchical features from the vehicle headlamp image, reducing computational complexity compared to traditional CNN systems. The fully connected layer is updated using backpropagation optimized with stochastic gradient descent to enhance the training procedure while maintaining discriminative properties. Experimental validation with a dataset of 13,300 training images and 2660 testing images demonstrates the robustness of the proposed method against various distortions. PCNN achieves superior performance, surpassing state-of-the-art techniques with an average accuracy of 99.51% over 38 car models using the PLUS dataset. The effectiveness of PCNN is further confirmed using the comp-cars dataset, achieving an accuracy of 89.83% over 357 vehicle models.

Lee et al. [10] proposed a deep learning approach using the SqueezeNet architecture for vehicle model recognition. The SqueezeNet architecture is modified to include bypass connections between fire modules, improving the efficiency of the MMR system. The authors collected a large-scale dataset of vehicle images, containing over 291,602 labeled images for training and evaluation. The proposed SqueezeNet-based model achieved a recognition rate of 96.3% at the rank-1 level with a fast processing time of 108.8 ms. The model's file size is compressed to less than 5 MB, making it viable for real-time applications. The study compared the proposed method with popular deep networks like AlexNet and GoogLeNet, demonstrating superior performance and efficiency. The residual SqueezeNet, with added bypass connections, further enhanced recognition accuracy without increasing the model size significantly. The proposed approach outperformed existing methods and achieved state-of-the-art rank-1 accuracy on a dataset of 766 vehicle models. The study provides insights into the design and evaluation of compact CNN architectures for efficient vehicle MMR.

With several techniques proposed for car model detection, the popular ResNet [11] models have not yet been investigated. Therefore, this study aims to explore popular ResNet variants, including ResNet18, ResNet50, and ResNet101, to assess their performance and compare their effectiveness in car model detection. The contribution of this study lies in testing ResNet models to evaluate the efficacy of transfer learning on these models for car model detection, an aspect that has not been examined previously.

3   Methodology

All the car model images were obtained from Google and Bing image searches. Each car model was searched, and the images were downloaded one by one. The downloaded car model images include Honda Accord, Honda BRV, Honda City, Honda Civic, Honda CRV, Honda CRZ, Honda HRV, Nissan Almera, Nissan Leaf, Nissan Navara, Nissan Xtrail, Perodua Alza, Perodua Aruz, Perodua Ativa, Perodua Axia, Perodua Bezza, Perodua Myvi, Perodua Viva, Proton Exora, Proton Iriz, Proton Perdana, Proton Persona, Proton Saga, Proton X50, Proton X70, Toyota Avanza, Toyota Camry, Toyota Fortuner, Toyota Harrier, Toyota Hilux, Toyota Innova, Toyota Rush, and Toyota Vios. In total, 33 classes are trained in the deep learning model. A total of 6730 car images were collected. 3229 of the images were used as a testing set. The remaining images are used for training the deep learning model. Of the remaining images, 30% were used as a validation set to prevent overfitting.

The images come in various sizes. Therefore, the images were resized to a size of 224 × 224 so that they can be used to train the deep learning model. ResNet101 was used for training. Transfer learning was performed. This was done by removing the final fully connected layer and then inserting a new one which contains only 33 classes. The model was then trained using stochastic gradient descent with a momentum optimizer, a mini-batch size of 128, a learning rate of 0.01, and a validation patience of 4. After the training was completed using the training set and the validation set, the model was tested on the testing set.

The testing set contains images that the model had never seen before. These images were used to test the model's performance. A confusion matrix was obtained by calculating the number of true positives, true negatives, false positives, and false negatives. These values were used to calculate the accuracy, average recall, and average precision of the model. The performance results were then compared to other models like ResNet18 and ResNet50 that were trained on the same dataset. Furthermore, the methods proposed in previously published papers were also implemented using the dataset utilized in this study for performance comparison. Figure 1 shows the flowchart of the methodology.

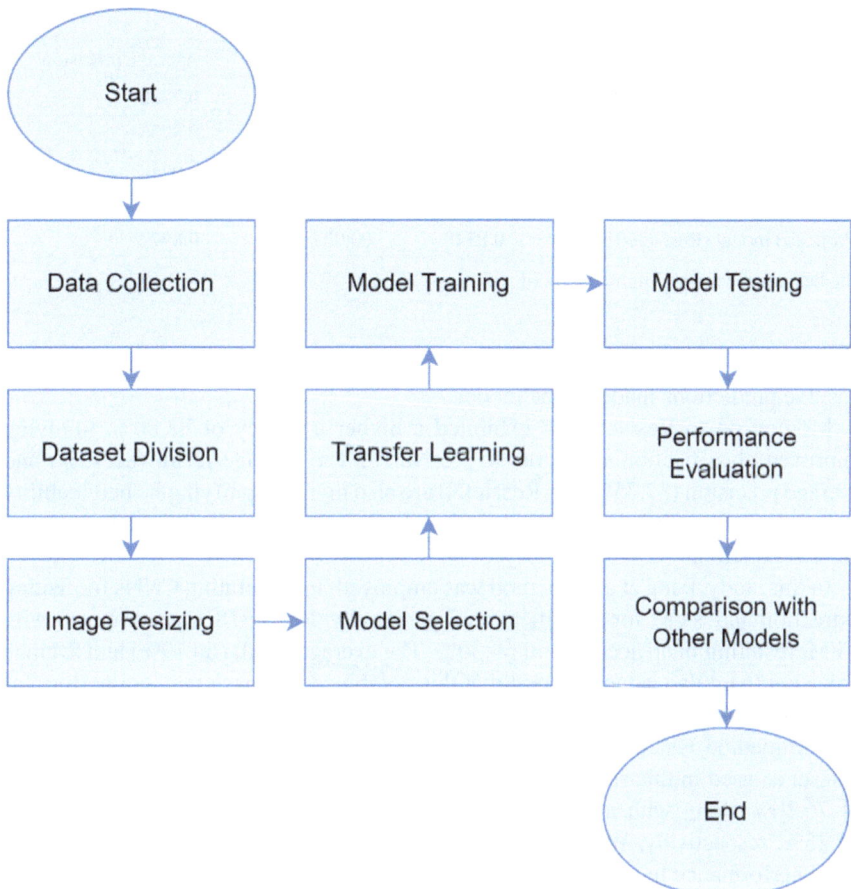

Fig. 1 Flowchart of the methodology

4 Results and Discussions

The results obtained from the proposed model (ResNet101) are discussed in this section and compared to other methods. The accuracy, average recall, and average precision were obtained from the methods. Overall, the results obtained from each method are presented in Table 1.

The table presents a comparative analysis of various methods for car model detection, focusing on accuracy, average recall, and average precision. These metrics are fundamental in evaluating the performance of machine learning models, particularly in computer vision tasks.

Starting with ResNet18, the model achieved an accuracy of 72.70%, indicating that it accurately identified the car models about 72.70% of the time. The average recall of 72.78% suggests that the model effectively identified true positives in relation to

Table 1 Results obtained from various methods

Method	Accuracy	Average recall	Average precision
ResNet18	0.7270	0.7278	0.7788
ResNet50	0.8280	**0.8275**	0.8735
Fang et al. [8] (ResNet18 + SVM)	0.6030	0.6019	0.6442
Lee et al. [10] (AlexNet)	0.7630	0.7628	0.8125
Proposed model (ResNet101)	**0.8340**	0.8077	**0.8922**

The best results are highlighted in bold

actual positives, while the average precision of 77.88% indicates the precision of the positive predictions made by the model.

Moving on to ResNet50, it exhibited a higher accuracy of 82.80%, implying improved classification compared to ResNet18. The average recall (82.75%) and average precision (87.35%) for ResNet50 are also higher, signifying its better ability to correctly identify true positives and make accurate positive predictions in car model detection.

In this study, Fang et al.'s method was employed, incorporating CNNs for feature extraction and SVM for classification. Specifically, ResNet18 was combined with SVM, resulting in an accuracy of 60.30%. The average recall (60.19%) and average precision (64.42%) values suggest that this combined approach may not perform as well as ResNet50, indicating room for improvement.

The method tested by Lee et al. is implemented in this study using the same dataset as used in this study. The model utilized is AlexNet, achieving an accuracy of 76.30% along with average recall and average precision values of 76.28 and 81.25%, respectively. While not as high as ResNet50, these values demonstrate a good performance in accurately identifying car models.

The proposed model using ResNet101 achieved the highest accuracy at 83.40%, highlighting its superior capability in correctly classifying car models. The average recall (80.77%) and average precision (89.22%) further emphasize the model's ability to identify true positives and make accurate positive predictions, making it a strong contender for car model detection.

In essence, the suggested model stands out as the optimal choice for car model detection. Utilizing a uniform dataset for training across all methods in the table ensures an equitable basis for performance evaluation. The proposed model, distinguished by its greater depth, excels in precise extraction of car model features compared to the alternative models assessed by various researchers.

5   Conclusions

In this study, an effective methodology for car model detection was developed, utilizing a dataset of 6730 car images spanning 33 distinct classes. The images underwent preprocessing and resizing to standardized dimensions of 224 × 224, enabling the utilization of ResNet101 architecture for transfer learning. Superior model performance was achieved through meticulous training and validation procedures, with stochastic gradient descent and momentum optimization being employed. The proposed model surpassed other methods, demonstrating an accuracy of 83.40%, along with the highest average recall (80.77%) and average precision (89.22%). The potential of ResNet101 in precise car model identification is emphasized by these compelling results, showcasing its applicability in real-world scenarios and endorsing the efficacy of advanced deep learning architectures in computer vision applications. Future research will further refine and extend these advancements in car model detection, promising enhanced performance and broader applications.

References

1. M.C.S. Tang, S.S. Teoh, H. Ibrahim, Z. Embong, Neovascularization detection and localization in fundus images using deep learning. Sensors **21**(16), 5327 (2021). https://doi.org/10.3390/s21165327
2. M.C.S. Tang, S.S. Teoh, H. Ibrahim, Z. Embong, A deep learning approach for the detection of neovascularization in fundus images using transfer learning. IEEE Access **10**, 20247–20258 (2022). https://doi.org/10.1109/ACCESS.2022.3151644
3. M.C.S. Tang, S.S. Teoh, Blood vessel segmentation in fundus images using hessian matrix for diabetic retinopathy detection, in *Proceedings of the 2020 11th IEEE Annual Information Technology, Electronics and Mobile Communication Conference (IEMCON)*, pp. 0728–0733 (2020). https://doi.org/10.1109/IEMCON51383.2020.9284931
4. M.C.S. Tang, S.S. Teoh, H. Ibrahim, Retinal vessel segmentation from fundus images using DeepLabv3+, in *Proceedings of the 2022 IEEE 18th International Colloquium on Signal Processing and Applications (CSPA)*, pp. 377–381 (2022). https://doi.org/10.1109/CSPA55076.2022.9781891
5. M.C.S. Tang, S.S. Teoh, Brain tumor detection from MRI images based on ResNet18, in *Proceedings of the 2023 6th International Conference on Information Systems and Computer Networks (ISCON)*, pp. 1–5 (2023). https://doi.org/10.1109/ISCON57294.2023.10112025
6. Z. Selmi, M. Ben Halima, A.M. Alimi, Deep learning system for automatic license plate detection and recognition, in *Proceedings of the 2017 14th IAPR International Conference on Document Analysis and Recognition (ICDAR)*, pp. 1132–1138 (2017). https://doi.org/10.1109/ICDAR.2017.187
7. A. Psyllos, C.N. Anagnostopoulos, E. Kayafas, Vehicle model recognition from frontal view image measurements. Comput. Stand. Interf. **33**(2), 142–151 (2011). https://doi.org/10.1016/j.csi.2010.06.005
8. J. Fang, Y. Zhou, Y. Yu, S. Du, Fine-grained vehicle model recognition using a coarse-to-fine convolutional neural network architecture. IEEE Trans. Intell. Transp. Syst. **18**(7), 1782–1792 (2017). https://doi.org/10.1109/TITS.2016.2620495
9. F.C. Soon, H.Y. Khaw, J.H. Chuah, J. Kanesan, PCANet-based convolutional neural network architecture for a vehicle model recognition system. IEEE Trans. Intell. Transp. Syst. **20**(2), 749–759 (2019). https://doi.org/10.1109/TITS.2018.2833620

10. H.J. Lee, I. Ullah, W. Wan, Y. Gao, Z. Fang, Real-time vehicle make and model recognition with the residual squeezenet architecture. Sensors **19**(5), 982 (2019). https://doi.org/10.3390/s19050982
11. K. He, X. Zhang, S. Ren, J. Sun, Deep residual learning for image recognition, in *Proceedings of the 2016 IEEE Conference on Computer Vision and Pattern Recognition (CVPR)*, pp. 770–778 (2016). https://doi.org/10.1109/CVPR.2016.90